AF711168

HOW TO CONTROL ORCHID VIRUSES: THE COMPLETE GUIDEBOOK

How To Control Orchid Viruses:
The Complete Guidebook

by Gail C. Wisler

MAUPIN HOUSE

HOW TO CONTROL ORCHID VIRUSES:
THE COMPLETE GUIDEBOOK
By Gail C. Wisler
Copyright 1989 by Gail C. Wisler
Produced by Ratzlaff & Associates
Published by Maupin House Publishers
P. O. Box 90148
Gainesville, FL 32607
All rights reserved. No part of this book may be reproduced, stored in a retrieval system, or transmitted, in any form or by any means, electronic, mechanical, photocopying, recording, or otherwise, without the written permission of the publisher.

Printed in the United States of America
Price: $19.95

Library of Congress Cataloging-in-Publication Data

Wisler, Gail C., 1956-
How to control orchid viruses.

Bibliography: p.
Includes index.
1. Orchids -- Diseases and pests --Control.
2. Plant viruses -- Control. I. Title.
SB608.065W57 635.9'3415 88-32575
ISBN 0-929895-00-2

CONTENTS

PREFACE

This book is long overdue. It is written for the quality-conscious orchid grower who wants a comprehensive, yet practical reference guide about orchid viruses. This book puts it all together for the orchid hobbyist and commercial grower who wants to look just one place for answers about a sometimes taboo subject.

Because the orchid industry is devoted to quality, viruses should be taken seriously. It is generally recognized throughout the industry that orchid viruses can cause substantial damage. Viruses are especially insidious since infected plants cannot be cured. Plants remain alive indefinitely as inferior specimens — despite the best of care. All the while they represent a danger to nearby healthy plants. Viruses are also primarily responsible for the finite life expectancy of varietal orchids. In fact, many of our earliest clones that hold both historical and esthetic value may be with us today were it not for viruses.

While the virus-infected plants may sometimes appear to grow normally, they never perform as well as their healthy counterparts. Since orchidists set quality standards unexcelled by most other plant growers, the fate of the unsuspecting orchidist who exhibits infected flowers may indeed be an unhappy one.

The susceptibility of orchids to virus diseases was conclusively established in the 1940s. Since then, technological advances in orchid virology have lead to a considerable body of knowledge about them. At least 25 different orchid viruses have been identified, and accurate methods to diagnose them have been developed.

Today, we have the technology to grow orchids free of all known viruses. Theoretically, researchers have given growers all the tools and knowledge needed to establish and maintain a virus-free orchid collection. But recent surveys show that despite the technology, there is high incidence of virus contamination in cultivated orchids, wherever they are grown throughout the world. No orchid collection — private or commercial — is likely to be completely virus-free.

But you do not have to live with viruses in your orchid collection. That's the message of this book, and the author, Gail Wisler, is eminently qualified to deliver it. A plant pathologist since 1974, she recognized the need for a scientifically accurate, comprehensive guidebook written especially for orchidists. This book successfully combines her scientific background with a thorough understanding of the grower's informational needs, based on years of first-hand experience.

It has been written that genius is the ability to reduce the complicated to the simple, and this was precisely her goal. Gail bridges the "jargon gap" that can separate scientists from growers.

I can vouch for the scientific accuracy of this book. I believe it is a valuable contribution to the orchid industry, and it provides the technical information necessary to control viruses in terms that the orchidist can understand

Dr. F. W. Zettler
Professor
Plant Pathology
University of Florida
Gainesville, Florida
October, 1988

INTRODUCTION

You Can Fight Viruses — and Win!

As an orchid grower, you are no doubt familiar with the subject of orchid viruses. These important and well-studied orchid diseases are a recurring theme found in much of the orchid literature you are likely to pick up.

Unfortunately, many hobbyists tend to put an orchid virus article down shortly after they begin reading it. Many articles are simply too technical or specialized. They seem to be written for the scientist — even though the intended audience was probably the hobbyist or small commercial grower.

Hobbyists are profoundly interested in their collections. Despite the interest, though, most do not have the time or desire to cut through the scientific literature for the answer to their questions. They need practical information, not a difficult-to-read treatise. They want a guidebook that helps them understand orchid viruses. They could profit from a how-to manual that shows them how viruses in their collections can be controlled.

This book does just that. It connects the well-intentioned scholar and the interested hobbyist who needs research-based, practical information. Its foundation is the common ground shared between the world of research and the concerned hobbyist: the specter of orchid viruses.

Viruses are feared by the orchid industry, and the fear is well-founded. Viruses have made their way into many of the world's most valuable and treasured collections. Viruses decrease a collection's overall value and productivity. Since many orchid clones are extremely valuable, this kind of loss can be devastating to a grower's investment — in both money and time.

It is bad enough when an orchid displays the symptoms associated with virus infection. That plant, and all others propagated from it, will always be infected. Unfortunately, viruses pack a double-whammy punch. Not only do they gradually destroy the quality of plants and blooms, but they can also invade a collection and lurk there without visible symptoms. Without telltale symptoms, a particular virus can go undetected. Yet even without symptoms they damage your plants, which then become a sort of orchid "Typhoid Mary."

The grower may unknowingly spread the virus throughout the collection by propagating the plant. The plant could be mericloned, in the process perhaps exposing other collections.

These Typhoid Marys can be hard to detect, and they have caused orchid growers to respond to viruses with varying levels of concern. Some growers are extremely concerned about orchid virus damage, regardless of whether a plant shows symptoms or not. For these growers, nothing short of indexing, or testing, their collections will do. Others, however, choose to ignore viruses. Without symptoms, they claim, what damage is done? Why bother to clean up a collection?

The answer is simple. Although an orchid virus may not express visible symptoms of damage all the time, it always causes some damage to the plant. A virus is incurable. It robs the infected plant of vigor and weakens it, making it vulnerable to secondary diseases caused by bacteria and fungi. And, of course, the weakened plant provides a source of infection for other healthy plants, thus threatening your collection by a back door route.

It is important for a grower to control viruses. And you can! Sound hopelessly optimistic? Not really. Viral diseases have been controlled in many other commercially important plants, including chrysanthemum, gladiolus, lettuce, and potato. Why not orchids?

The first step toward control is a recognition of the problem. Do not bury your head in the sand and pretend you can live with viruses because you have given up trying to control, or even understand them. You can fight viruses — and win!

The second important step in successful anti-virus campaigns is for growers to search for the best and most accessible indexing system and control recommendations. Establish the procedures in your own operations, and expect industry to follow the same standards of sanitation.

The final step is for the grower — as consumer — to strongly demand the highest possible production standards from the supplier, in this case, the orchid industry at large. All your efforts and expense as an individual grower will be in vain if you fail to demand the highest quality plants from commercial growers.

Virus control can be achieved in the orchid industry as well, especially when you consider that the two most important and widespread orchid viruses, cymbidium mosaic (CyMV) and odontoglossum ringspot (ORSV), are spread only by growers and their propagating instruments. No other vector of these two viruses, or other members of their virus groups, has ever been proven!

One Grower's Success Story

Dr. Tom Sproles, an orchid hobbyist in New Orleans, Louisiana, was determined to clean up his collection of 30 years. In 1978, Sproles tested his entire collection for CyMV and ORSV. At that time, the collection showed 100 percent infection by one or both of these viruses.

With the help of myself and other virologists at the University of Florida, Sproles rebuilt his collection over the next four years. A retest in 1982 showed he had reduced viruses to 3.4 percent. A second test in 1985 indicated virus incidence at 0.17 percent. All but one of 580 samples tested free of either virus (34, 35).

Today, Sproles says the effort was worth it.

"It's hard to evaluate whether my plants are better now because I don't have anything to compare it with since I've been virus-free for several years," he says. "All I can say is that all the plants now bloom and flower regularly. Before, they didn't. And they appear more thrifty. I feel better as a grower that I'm not spinning my wheels and wasting time on infected plants."

Qualitative differences are admittedly hard to measure, yet

Sproles can remember a time when viruses robbed his plants of vigor and caused irregular blooming. Sproles' experiences can be yours. This book outlines the same principles that he used to eliminate viruses from his collection.

First, he recognized the problem. Then he selected a testing method, which for him was the serological technique of immunodiffusion. Sproles chose to destroy his infected plants — a final solution that this book does not insist on, by the way. Infected plants are, after all, not totally worthless. With proper care they can be used for breeding purposes and for use as positive controls in diagnostic tests.

Sproles then positioned his plants in the greenhouse so that water from one orchid would not drip on another. No pot was used twice. Wire hangers were flame-sterilized to glowing. All divisions were made with a new razor blade. Snails and insects were kept to a minimum. Unthrifty plants were isolated and retested, and all new plants were quarantined and tested before joining the collection.

At first reading, Sproles' procedures may seem extreme or difficult. But they worked for him. Your efforts — however far you choose to go — will also pay off. You will protect your valuable investment and increase the overall quality of your crop, as well as contribute to the health of the industry worldwide.

Looking Toward the Future

The future prospects for orchid virus control look good. With good grower education and sanitation, CyMV and ORSV can be effectively eliminated. Over time, detection and control methods will improve. Growers are increasingly aware that a virus-free collection is a practical and attainable goal.

I hope you will allow this book to guide you to a better understanding of viruses. Let it alert you to your role as an active participant in the process of virus control. The book explains the various virus indexing systems and helps you choose which one suits your collection best. It takes you step-by-step through two diagnostic techniques and thoroughly explains another used by some laboratories. It shows you ways to control insects and advises you on sanitation for tools and pots. Finally, it can guide you toward complete control so you can eliminate, or at the very least, minimize viral infections in your collection.

Because of the crossover of interest between orchid hobbyists and gesneriad growers, a section in the appendix addresses the virus situation in that interesting plant family. The appendix also includes a list of suppliers, and references for further information.

If you learn nothing else from this book, I hope you will carry away the message that you as a collector are not helpless to control orchid viruses. Your collection can be as clean from viruses as you choose it to be.

Gail C. Wisler
October, 1988

Chapter 1

INTRODUCING THE VIRUSES

Although some orchid viruses can be spread by an *insect vector*, they are spread more often by orchid enthusiasts themselves through *manual transmission*, or contaminated potting media and cutting tools. No matter how viruses spread from plant to plant, every experienced orchid grower is well aware of the devastating symptoms of flower color break and flower necrosis, both signs of orchid virus disease.

Manual Transmission

Typical symptoms of flower color break are caused by odontoglossum ringspot virus (ORSV), also known as TMV-O. Flower necrosis is caused by cymbidium mosaic virus (CyMV). These are the only two viruses that are exclusively transmitted by the tools of the orchid lover's trade. All research in this area indicates this to be true.

These two manually transmitted viruses are definitely what you would call survivors. In fact, they are so stable that they can survive on cutting tools, pots and even on potting media where orchids have been grown.

Virus particles are contained within the plant cells and are released when a plant is wounded by a tool used in flower harvesting or plant divisions. If the tool is not sterilized before you use it on your next orchid, you will very likely transmit it to the next plant you cut.

Basic Points to Remember

- CyMV and ORSV are usually considered together because they are so easy to transmit.
- They have worldwide distribution. They are found in almost every orchid collection and in many orchid genera.
- They do not always produce characteristic symptoms. That means that the virus may be infecting your plant but sometimes symptoms may not be present at all, or they may resemble some other, non-viral disease.
- Interestingly enough, even though you can infect your own plants, there are no conclusive tests that have proven these two viruses are spread either through seed, by insects, or other agents. Various studies have supported the fact that CyMV and ORSV are not transmitted through seed, since neither seedlings nor wild, seedborne species have been found to be infected with these viruses (39, 41, 43).

There is a slight possibility of mother-daughter transmission, however, in seedlings derived from the green-pod technique of flasking if the mother plant is infected with either CyMV or ORSV. This infection can apparently happen through contact with the parent tissue within the flask. One study, for example (42), showed that one out of 123 *Dendrobium* seedlings derived from green-pod culture was infected with CyMV. As you can see, infection in this way may be possible, but is extremely rare. This subject is covered in greater detail in Chapter 3.

Orchid Categories and Their Susceptibility to Virus Infection

Orchids are grown for cut flowers or potted plants, and they are produced throughout the world in a variety of climates and habitats. They are propagated from seed, either naturally in the wild or as hybrid crosses, and vegetatively through cuttings or meristem culture *in vitro*. The major virus offenders of cultivated orchids are CyMV and ORSV. How often they occur is directly related to how often an orchid is handled with tools or repotted. Studies indicate that the longer the flower or the plant

TABLE 1. Incidence of cymbidium mosaic and odontoglossum ringspot viruses detected by G.C. Wisler in individual collections of cultivated orchids between 1980 and 1985.

Genus	Number of Plants Tested Infected/Total	Total %
Angraecum	2/51	4
Ascocenda	14/81	17
Brassocattleya, Brassolaeliocattleya	147/487	30
Cattleya	123/496	25
Cymbidium	100/204	49
Dendrobium	51/194	26
Epidendrum	19/161	12
Laelia	9/69	13
Laeliocattleya	151/431	35
Miltonia	2/30	7
Odontoglossum	3/25	12
Oncidium	12/147	8
Paphiopedilum	6/168	4
Phalaenopsis	74/267	28
Potinara	23/54	43
Sophrolaeliocattleya	37/151	25
Vanda	30/151	20
Miscellaneous	59/363	16

is in a collection, the greater the chance it will be virus-infected.

CyMV was first described in *Cymbidium* and ORSV in *Odontoglossum*. But this does not mean that each of the viruses restricts itself to either genus. Both have wide host ranges within the Orchidaceae, the world's largest and most diverse plant family (Table 1), with 800 genera, 25,000 species and at least as many hybrids (30, 32). Surveys taken of collections around the world show that CyMV and ORSV are present in almost every collection to some degree. How much infection largely depends on which type of orchid is grown and the production habits of the grower. Certain types of orchids, such as the *Cattleya* alliance, are particularly notorious for virus infection, while others, such as botanicals, are not. The following gives you some general guidelines about the risk of infection in various orchid categories.

Large Flowering Plants These orchids produce showy flowers and are highly valued in the industry, especially as cut blooms. Some of these are the *Cattleya* alliance, *Cymbidium*, *Phalaenopsis*, *Dendrobium*, *Vanda*, and *Oncidium*. These orchids are often propagated by standard means with pruning tools and through *in vitro* meristem culture. Their flowers are also harvested frequently and are cut with knives or razor blades. *They show a high incidence of virus infection.*

Botanicals Although they are usually extremely small, botanicals have other characteristics that make them interesting to the grower. The flowers are often minuscule, with plants remaining small even after many years of growth. *Pleurothallis*, *Bulbophyllum* and *Lepanthes* are examples of this group. They are rarely propagated, and their flowers are seldom harvested. *Because of that, they have a low incidence of virus infection.*

Infected orchids of this type can be found, however, so they too are susceptible.

Wild Orchids This category includes a wide variety of physical types, including extremely small orchids and some very showy ones. The common characteristic is that all have been propagated from seed while in their native habitat. Until they were collected from the wild, no tools had been used to make cuttings or harvest flowers. *Wild orchids have virtually no incidence of manually transmitted virus infection.*

Regardless of flower type, orchids collected wild from their native habitat are free of CyMV and ORSV (2,39,43) because these two viruses are not seedborne. Like every other orchid type, wild orchids are, however, susceptible to viruses. Surveys show that the longer the wild orchid is in cultivation the higher the incidence of infection (36). Again, the risk comes from handling for propagation or flower harvest.

Hybrid Seedlings These orchids are produced in cultivation. There are two variations in how the pods are handled. Once the seed pod develops from a hand-pollinated flower, the pod is either allowed to dry and burst naturally as in nature, or is harvested prematurely and placed in a sterile flask of nutrient media where it develops *in vitro*. Once these plants grow to an appropriate size, they are subjected to the same propagation practices as any in cultivation. *Virus incidence is very low, but can occur depending on the treatment of the seed pod.* This subject will be discussed more thoroughly in Chapter 3.

Hybrid seedlings are originally free from CyMV and ORSV, but as they grow to a size suitable for dividing or flower harvesting, they too can become infected by the tools of the orchid trade.

CyMV is in almost all genera and in all orchid collections and is much more prevalent than ORSV. Depending on the collection, CyMV infection generally appears in about one in five orchids. In contrast, ORSV occurs in an average of two out of 100 orchids. The only exception is with cool weather cymbidiums, where the percentages are usually reversed (16, 18). In this genus, for some reason not yet understood, ORSV occurs most often.

Cymbidium ringspot virus is another virus for which there is no known insect vector. It is mentioned here as a caution for growers of terrestrial orchids. Researchers in England have reported this virus in cymbidiums. It is apparently spread through soil from one plant to another. Growers of terrestrial orchids should be aware of this danger and take care to use sterile potting media.

Insect-Transmitted Viruses

Manually transmitted orchid viruses affect a large percentage of orchids and cause a significant amount of economic damage. Because of that, researchers have placed a great deal of effort in diagnosis and control programs for CYMV and ORSV (27, 36, 38, 40). In contrast, much less is known about some of the other viruses that are transmitted by various insects.

Contaminated cutting tools do not play an important role in transmitting this type of infection. These viruses are less stable and are less likely to survive outside the plant cell. They usually need a vector, which in most cases is an insect.

Points to Remember

- Little is known about insect-transmitted orchid viruses at this time, and they may have limited distribution. For example, cucumber mosaic virus (CMV) has been reported in

Taiwan; bean yellow mosaic (BYMV) in Japan and the United States; and the vanilla mosaic virus (VMV) in French Polynesia (26, 37).

● There is less information available about insect-transmitted viruses than viruses which are manually spread. Because they are so widespread, CyMV and ORSV have received a great deal of attention. So much so, in fact, that the effects of insect-transmitted viruses have not been as well documented.

● In most cases, symptoms are characteristic, that is, they reveal the presence of the virus.

● The infection can come into your collection from alternate plant hosts such as weeds which also serve as a host for the insect.

The aphid is the main culprit for transmitting these orchid viruses (46, 47). Table 2 lists most of the orchid viruses that have been described, their probable virus group and likely natural vectors. As you can see, the aphid is the most notorious virus vector. Although aphids rarely colonize orchids extensively, they are often found on inflorescences and buds. Because of this, you as an orchid grower may see evidence of these "other" viruses as eruptions on a localized area of your collection.

These eruptions have been observed for *dendrobium vein necrosis virus* (DVNV)(15, 28), a member of the closterovirus group whose members are known to be transmitted by aphids. Fortunately for growers who are looking for DVNV, the virus produces characteristic symptoms that reflect its name, namely, necrotic flecking along the veins on both flowers and foliage of infected dendrobiums.

Additional examples of viruses that are likely transmitted by aphids include *bean yellow mosaic (BYMV), turnip mosaic (TuMV), clover yellow mosaic (CYVV), cucumber mosaic (CMV),* (22) *dendrobium mosaic*, several rhabdoviruses including *orchid fleck* (8, 14), and the recently described *Vanilla potyvirus* (37). Little is known about the distribution and host range of these infrequently reported and diagnosed viruses within the Orchidaceae.

For orchid growers, routine diagnostic tests have not been extensively developed for these "other" orchid viruses. Re-

TABLE 2. Probable taxonomic affinities and natural means of transmission of 23 orchid viruses.

Virus	Group*	Natural vector*
Bean yellow mosaic	potyvirus	aphids
Brazilian bacilliform	(rhabdovirus)	(aphids)
Clover yellow vein	potyvirus	aphids
Cucumber mosaic	cucumovirus	aphids
Cymbidium mild mosaic	(rhabdovirus)	(aphids)
Cymbidium mosaic	potexvirus	unknown
Dendrobium rhabdovirus	rhabdovirus	(aphids)
Dendrobium mosaic	potyvirus	aphids
Dendrobium vein necrosis	(closterovirus)	(aphids)
Filamentous cypripedium	(potyvirus)	(aphids)
Filamentous orchid	unknown	unknown
Grammatophyllum bacilliform	(rhabdovirus)	(aphids)
Isometric masdevallia	unknown	unknown
Laelia red leafspot	rhabdovirus	(aphids)
Long orchid rhabdovirus	(rhabdovirus)	(aphids)
Odontoglossum ringspot	tobamovirus	unknown
Orchid fleck	rhabdovirus	(aphids)
Phalaenopsis chlorotic spot	rhabdovirus	(aphids)
Short orchid rhabdovirus	(rhabdovirus)	(aphids)
Tobacco rattle	tobravirus	nematodes
Tomato ringspot	nepovirus	nematodes
Turnip mosaic	potyvirus	aphids
Vanilla potyvirus	(potyvirus)	aphids

*Groups in parentheses are tentative and not officially acknowledged by the International Committee for the Taxonomy of Viruses (Matthews, 1982). Vectors in parentheses also are tentative and not yet established conclusively.

cently, however, light microscopic diagnosis by viral inclusion bodies has made it possible to recognize the viruses (10). For example, characteristic DVNV inclusion bodies have been observed in the phloem cells of infected plants with the light microscope.

Several different *rhabdoviruses* have also been reported in orchids. The identity of these viruses is still largely a mystery, but they are assigned to the rhabdovirus group because of their characteristic bullet — or bacilliform — shape revealed under an electron microscope, and the production of plate-like nuclear inclusion bodies seen with the light microscope.

Although these viruses can be transmitted by aphids, leafhoppers, or plant hoppers, aphids are the only known rhabdoviral transmitter on orchids. Symptoms of rhabdoviruses are fairly characteristic in some of the orchid hosts. Often, on *Gongora*, *Brassia* and *Cochleanthes*, symptoms appear as sunken, necrotic, concentric ringspots, or a "bulls-eye" type pattern which can often be extremely irregular in shape.

Cucumber mosaic (CMV) has been reported from Taiwan on *Phalaenopsis* (26). This aphid-transmitted virus has a wide host range outside the Orchidaceae. Because of that, you should be careful to eliminate alternate sources of inoculum from surrounding areas where common plant hosts may be infected with CMV, such as *Commelina* sp. (wandering jew); Zinnia, spinach, *Vigna unguiculata* (cowpea); *Nicotiana tabacum* (tobacco). The characteristic white streaking and distortion symptoms of CMV in infected flowers simplify diagnosis (see back cover).

Recently, a virus has been discovered on *Vanilla* in French Polynesia (37). The virus is suspected to be transmitted by aphids although no evidence of natural spread was observed first-hand in the *Vanilla* plantations. The virus was experimentally transmitted in the laboratory by aphids, however. In the field it appeared that workers unwittingly propagated infected *Vanilla* vines without recognizing the symptoms. The symptoms, however, are characteristic and easy to spot, and in this case, the virus is controlled simply by careful roguing. It is not known at this point if the virus infects other orchid genera.

Important Viruses of Orchids

Now that you are familiar with the means by which orchid viruses are transmitted and the likely risks of infection in various categories of orchids, you should know about the important viruses you may encounter.

Cymbidium mosaic virus (CyMV) This virus is probably the most prevalent and widely distributed orchid virus, found throughout the world in most genera tested. CyMV has been found in cultivated orchids only, not in those collected from the wild. It is transmitted only through contaminated tools, pots, etc., not by seed or insects. Symptoms are not always expressed, so that another means of diagnosis is needed. Indicator plants and serological techniques are excellent diagnostic tools.

Odontoglossum ringspot virus (ORSV) Like CyMV, this virus is also commonly found in cultivated orchids but never in wild orchids. Incidence of ORSV is usually not as high as CyMV, except in cool-weather cymbidiums in which ORSV incidence is very high. ORSV is transmitted manually, and not by aphids or through seed. As with CyMV, symptoms are not always expressed, so a diagnostic method such as indicator plants or a serological technique is recommended.

Cymbidium ringspot virus (CRSV) This virus has a limited distribution, found only in England in terrestrial cymbidiums. Transmission in nature occurs through the soil, so there may be a risk for terrestrial orchids grown without sterile potting media. Symptoms and indicator plants may be the best means of identification.

Cucumber mosaic virus (CMV) CMV has been reported in Taiwan in *Phalaenopsis*. This virus is aphid-transmitted and infects a wide range of plants outside the Orchidaceae. Its distribution to other parts of the world and to other orchid genera is not generally known. Symptoms, indicator plants, light and electron microscopy are all feasible diagnostic techniques for CMV.

Rhabdoviruses Although their exact identities remain obscure, several rhabdoviruses have been reported, some from West Germany and Japan. One which has been characterized is orchid fleck virus (OFV). Rhabdoviruses have been reported in *Dendrobium*, *Cymbidium*, *Phalaenopsis*, and *Cochleanthes*. Aphids are the likely transmitter. Symptoms are likely to be expressed in most cases. Indicator plants, light and electron microscopy are good diagnostic techniques.

Bean Yellow Mosaic Virus (BYMV) BYMV has been reported in *Masdevallia* in West Germany, the United States and Japan. It is aphid-transmitted and, like CMV, has a wide host range outside the Orchidaceae. It is possible that BYMV could have come from wild orchids. Symptoms, indicator plants, serology and light microscopy are good diagnostic tools for BYMV.

Vanilla Mosaic Virus (VMV) VMV has been reported only in *Vanilla* found in French Polynesia. Aphids and propagation of infected vines are the probable means of transmission. At this time, other plant hosts are unknown outside of the genus *Vanilla*. Symptoms are an excellent means of diagnosis, but light and electron microscopy can also be used.

Tomato Ringspot Virus (TRSV) TRSV has been reported only in cymbidiums in the United States. It appears to have a limited distribution. Nematodes are known to transmit TRSV, so there may be a risk for terrestrial orchids grown in the absence of sterile potting media. Indicator plants are a useful diagnostic technique.

Chapter 2

TECHNIQUES OF ORCHID VIRUS DIAGNOSIS

Now that you have been introduced to the viruses infecting orchids, it is important for you to evaluate the situation in your own collection so that you can choose an appropriate diagnostic technique. Plants are diagnosed to discover if the orchids are infected, and if so, what viruses are present.

This chapter contains diagnostic techniques, including recognition through symptoms, indicator plants, microscopic and serological methods. Some are simple and inexpensive and can be easily performed by an individual grower. Other methods are more technically difficult and costly and should be performed only by trained personnel. Even though you may not be able to perform them yourself, it's still important to understand them since a thorough understanding of these methods will allow you as a grower to choose the most appropriate technique for your particular situation.

RECOGNIZING THE SYMPTOMS

Most viruses cannot survive for long periods of time outside a living plant cell. That is why viruses rarely, if ever, kill the plant they infect. Instead, the viral invader weakens its host by taking over the plant's metabolism. Because of this take-over effect, the most universal virus symptoms are a reduction in plant growth and flower production.

In some field-grown plants, such as potato, lettuce or chry-

Fig. 1 Individual leaves from two different *Cattleya* plants, both of which are virus-infected. One appears healthy while the other shows severe brown necrotic spotting, typical of cucumber mosaic virus (CyMV). Growers should remember that the healthy appearance of an orchid is no guarantee that the plant is not infected by a virus.

santhemum, yield reduction can be easily measured within one growing season. Usually, the yield reductions are significant enough to warrant grower-initiated virus certification programs. Unfortunately, many orchids grow slowly so that yield reduction from viral infection cannot be easily measured in a single growing season. Instead, successful production is often judged more by the quality of blooms — a subjective measurement — rather than by quantity.

Viral symptoms within the Orchidaceae are as diverse as the plant family itself. They can be *unexpressed*, which means that although viruses are present in the plant they cause no visible symptoms. Or the symptoms can be *expressed* and range from only slightly noticeable to completely unacceptable distortion and necrosis (Fig. 1).

Because of the genetic diversity of the Orchidaceae, viral symptoms are highly variable, even from plant to plant. Each clone is an individual and may respond differently to virus infection. A normal-looking plant may give rise to some very symptomatic characteristics in flowers, or they might not. Thus, you will not always be able to predict how viruses will be expressed in your orchids.

With some experience, a grower can learn to recognize symptoms fairly well if they are expressed. However, the real problem lies with the symptomless orchids, or *carriers*. Researchers do not know why some clones show severe symptoms while others that are also infected appear healthy. This fact alone justifies a virus testing program to identify carriers in a collection. It is important to emphasize that whether symptoms are expressed or not, an infected orchid can spread contamination throughout a collection.

Viral symptoms can be confused with other diseases or physiological disorders as well. Many of the symptoms produced are chlorotic and/or necrotic spots or streaks. Fungal and bacterial diseases, and physiological problems such as light intensity, fertilizer burn, and water imbalance can also cause similar symptoms. With some experience, you can distinguish these from signs of true viral symptoms. This section will provide guidelines and photographs that should help you recognize some of the textbook symptoms you may encounter.

Fig. 2 Mosaic pattern of light and dark green seen on the foliage of *Phaius tankervillea.*

CyMV

When expressed, symptoms of CyMV on foliage range from streaks or spots that can be *necrotic* (brown or black) and/or *chlorotic* (yellow). The yellow streaks can appear as a pattern of light and dark green known as a *mosaic*, or mottle. The patterns usually appear on both the upper and lower leaf surfaces (Fig. 2). This symptom is seen in some CyMV-infected orchids, particularly thick-leaved cattleyas.

Often, the necrotic spots are sunken and sometimes appear as a ring of necrotic tissue. Leaf symptoms such as these are seen in many orchid genera (Fig. 3).

Flower necrosis induced by CyMV is a real problem for orchid growers who sell primarily cut flowers, or for those who display blooming cattleyas at orchid shows. The notorious brown streaks of the sepals and petals appear several days after a *Cattleya* flower opens (4). The streaks usually begin along the midrib or large veins and coalesce, sometimes covering most of the flower area (see photograph on back cover). This varies according to the clone and cultural conditions. Severely affected plants may drop leaves prematurely.

Aside from cattleyas, flower symptoms are usually not expressed in many other orchid genera. *Phalaenopsis* flowers, for example, may show only a slight puckering of the sepals and petals or may show necrotic streaks on older flowers (4). Most often, however, CyMV infection can be noticed on a wide range of orchid genera by the characteristic necrosis, mosaic or mottle patterns on the foliage.

CyMV is the most prominent virus of most genera and in most collections. This is true except for the cool-weather cymbidiums grown in Australia and Holland where a higher incidence of ORSV has been reported (16, 18). The reason for this difference, whether genetic or regional, is not known. Another exception is the genus *Paphiopedilum*. Judging from the categories of risk discussed earlier, you would expect that *Paphiopedilum* would have a fairly high virus incidence because of the popularity of their flowers. Only one percent of all 'Paphs' sampled, however, was infected with CyMV. After discussing this phenomenon with other orchid researchers, we found

Fig. 3 Necrotic brown or black spotting and streaking are typical of a viral infection, in this case with CyMV.

growers rarely use tools to make divisions in Paphs. Instead, plantlets of this orchid are simply pulled apart by hand, thus avoiding contact with contaminated tools. With *Paphiopedilum*, infection from both CyMV and ORSV is extremely unusual, especially for a large flowering type.

Aside from these two exceptions, CyMV is the most prevalent virus in the *Cattleya* alliance, *Phalaenopsis*, *Cymbidium*, *Oncidium*, and *Epidendrum*, and any other type of showy orchid.

ORSV

ORSV, also known as TMV-O, also causes severely debilitating symptoms in a wide variety of orchid genera. This virus is responsible for *color break* symptoms often seen in cattleyas. Color break is a variegation in the natural solid color of a flower, most conspicuous in the lavender shades (see front cover). Severity of symptoms can vary from flower to flower. They become evident when the flower opens. In addition to the variegated pattern, affected flowers are often distorted in shape. ORSV color break symptoms and distortion can sometimes be mistaken for genetic disorders seen in orchid flowers. In this case, only a diagnostic test will definitively reveal the difference.

Obviously, an orchid that shows color break, or any virus symptoms for that matter, is of little value besides breeding under carefully controlled conditions. In contrast to fungal and bacterial diseases, removing tissue that shows the symptoms does not help eliminate the virus from the system of the infected plant.

Leaf symptoms of ORSV include sunken chlorotic and/or necrotic spots or rings similar to those caused by CyMV. These spots may appear slightly red in the purple-flowered *Cattleya* types.

Mixed Infections of CyMV and ORSV

Unhappily, mixed infections of these two viruses are common, especially in large, showy *Cattleya* clones. Symptoms of mixed infections may be more severe than with single infec-

Fig. 4 Bulls-eye pattern of necrosis is caused by a rhabdovirus on this *Cochleanthes*.

tions, with flower breaking likely to be extremely severe. The stunting effect and lack of flower production may also be more pronounced. Combinations of other viruses are possible, too, of course, but the CyMV and ORSV combination is the one most often seen.

CMV

Cucumber mosaic infections have been reported in some *Phalaenopsis* from Taiwan (26). Symptoms are obvious, but because the flowers are distorted with white and irregular streaking, the grower can easily mistake CMV for a genetic defect (see photo on back cover).

Fortunately, this virus is uncommon in the United States. This aphid-transmitted virus does pose a risk, however, because CMV has a very wide host range that even includes several weeds, such as *Commelina* sp. (wandering jew). For this reason, it is important to keep CMV — and its aphid vector — out of your collection in the first place. Once introduced, it could pose, under the right conditions, a serious threat to the entire collection.

If you recognize CMV symptoms, you should contact a diagnostician capable of detecting CMV as soon as possible. If CMV is detected, the plant and any propagations and divisions should be destroyed. Because CMV is transmitted by aphids, any colonies observed should be sprayed immediately. Insect control will be discussed further in Chapter 3.

Rhabdoviruses

Symptoms produced by rhabdoviruses can be highly variable, but as with the other viruses they can cause chlorotic and necrotic spots on leaves. In *Gongora* and *Cochleanthes*, these symptoms appear as irregular concentric ringspots that are necrotic (Fig. 4). Severely affected leaves may die and drop prematurely.

Floral symptoms have not yet been reported for orchid rhabdoviruses. It is also true, however, that relatively less research has been conducted in this area compared to CyMV

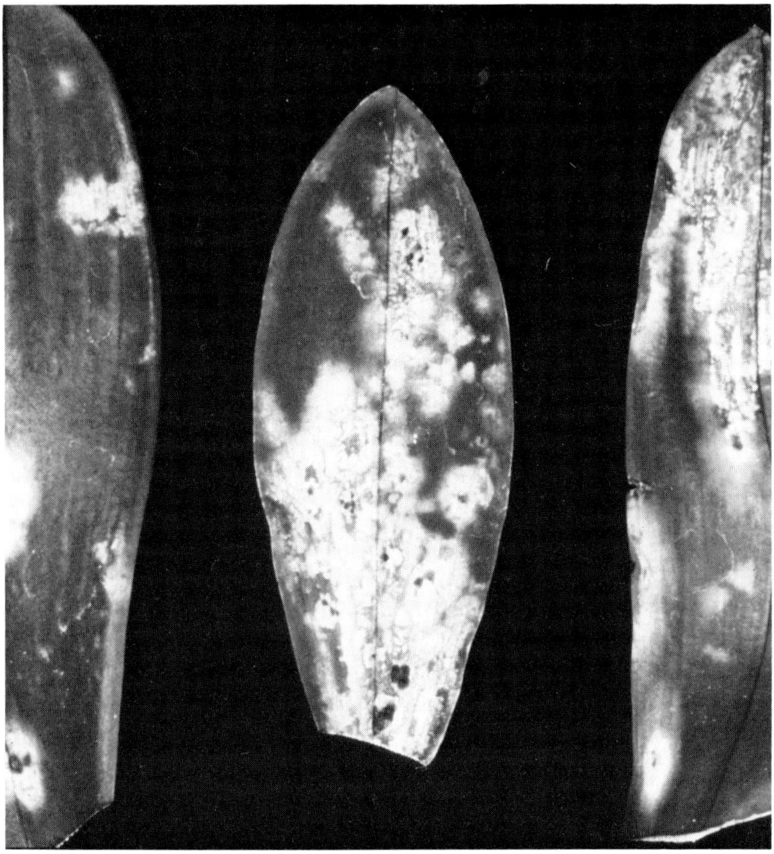

Fig. 5 Symptoms of orchid fleck virus, a rhabdovirus, in *Dendrobium nobile*. Note the display of sunken, parchment-like necrotic spots and patches. (Photo courtesy D. E. Lesemann)

and ORSV. With time, orchid virologists may be able to better recognize the diversity of symptoms produced by rhabdoviruses and learn more about their diagnosis. Aphids are also considered to be a transmitter here, so colonies should be controlled.

Orchid fleck virus (OFV), one of the rhabdoviruses, was first described in *Dendrobium*, *Cymbidium*, and *Phalaenopsis* from Japan and West Germany (26). It causes chlorotic and necrotic ringspots or flecks in leaves (Fig. 5).

Bean Yellow Mosaic Virus

BYMV produces mosaic and mottle patterns and sunken necrotic leaf spots. Like CMV, BYMV has a wide host range, including gladiolus, bean and clover, and is aphid-transmitted. Since some orchid hosts, such as *Masdevallia*, are terrestrial, it has been suggested that BYMV infection may have originally come from wild orchids. BYMV was first reported in Japan, but more recently in orchids exported to Germany from the United States (26).

Dendrobium Vein Necrosis Virus

DVNV was first reported in 1977 (15,28) in West Germany and later in Florida. It causes sunken patterns along the veins of leaves and flowers of dendrobiums. It is not known at this time if the virus infects other orchid genera.

The sunken areas can become necrotic, with patterns most noticeable on purple-flowered dendrobiums (Fig. 6) and difficult to see on white-flowered ones (15). Flowers fail to open fully. Aphids are the suspected vector of this virus, based on knowledge of other members of the closterovirus group, to which DVNV belongs.

Observe plants closely for early development of aphid colonies, particularly on inflorescences, and look for the characteristic symptoms.

Fig. 6 Symptoms of dendrobium vein necrosis virus (DVNV). Note the sunken pattern of pale necrotic tissue along the veins of the flowers.

Vanilla Mosaic Virus

This relatively "new" virus was described in 1987 by the author in French Polynesia. The characteristic mosaic and distortion of infected leaves betrays its presence and was responsible for a marked decrease in vine growth (37). Surveys conducted on the four major production islands showed that symptoms are reliable and easy to detect, with every leaf formed on an infected plant showing symptoms (Fig. 7). This has made the control of this virus an easy task for French Polynesian *Vanilla* growers and specialists. It is not known if this virus infects other orchid genera.

Tomato Ringspot Virus

This virus has been reported in *Cymbidium* Snowbird 'Jayhurst' (17). It produces faint chlorotic and necrotic streaks and flecks along leaf veins. Symptoms on flowers have not been observed. Early development of foliar symptoms are said to be easily confused with infections of CyMV or ORSV. This virus appears to be restricted to cymbidiums.

Cymbidium Ringspot Virus

First described in *Cymbidium* plants, this virus is also known to infect white clover (21). It is transmitted through the soil from materials released by the roots into the soil environment. Found in England, this virus is apparently not widespread. Symptoms include chlorotic ring-like mottle. When found in combination with CyMV, symptoms are reportedly intensified.

Virus Look-Alikes

Orchid virus symptoms can be mimicked, either through cultural conditions or fungal and bacterial infections. As you have seen, the most frequent symptoms of most viruses are necrotic or chlorotic spots and streaks.

However, viral-like symptoms can be caused by other means, too. For example, nutritional deficiency, excess fertil-

Fig. 7 Distortion and mosaic symptoms of light and dark green are typical of the *Vanilla* potyvirus.

izer, insect damage, sunburn, air pollution, chilling, pesticide spray damage, and specific fungal or bacterial disease can also cause you to mistakenly believe you have a viral infection in your collection.

Discovering the true cause can be confusing. The best way to avoid the task of eliminating the mimics is to follow accepted production practices. Fortunately, viruses usually leave telltale clues to help you make your diagnosis. When in doubt, consult a plant diagnostician.

Here are a few of the most common virus mimics.

Tip burn can be caused by excess fertilizer, lack of sufficient water or even root rot caused by soil fungi. A localized necrosis on the tips of the leaves is a telltale sign (Fig.8).

Mesophyll collapse is caused by chilling or cold water on leaves, or damage from insecticides (32). Newly expanding leaves are most susceptible. Damage appears up to several weeks later as sunken necrotic patches or spots. The distinction here is that with mesophyll collapse, subsequent growth will be normal.

Botrytis petal blight This fungal petal blight has been mistaken for CyMV. Botrytis usually shows up in cool-damp weather with insufficient air circulation. Unlike CyMV which expresses itself as streaks, Botrytis necrosis leaves discrete, pinpoint lesions scattered on the flowers.

Sunburn causes the yellow mottling pattern, or a dry, parchment-like appearance on leaves, which can sometimes look like CyMV. Severe scale insect infestation on the underside of the leaf can also appear this way. Fortunately, the sunburn can usually be associated with overexposure to intense sunlight, clearing up in subsequent growth when the plant is protected. Insects can, of course, be easily detected and controlled.

Fungal leafspots can be caused by a variety of fungi, including *Phyllostictina cactorum* and *Cercospora* sp. These disorders are also characteristic and can be diagnosed relatively easily (4,6). The necrosis often expressed by many of these diseases could be mistaken for that caused by a number of viruses. However,

Fig. 8 Tip burn on a *Cattleya* is a physiological condition. The location of the necrosis is a clue that it is not viral in origin.

treatment with the appropriate fungicide should remedy the problem. Contact your local agricultural agent or nursery for current pesticide recommendations.

Bacterial leafspots are caused by *Pseudomonas cattleyae* or *Erwinia carotovora* (4). A softer, watersoaked appearance of lesions distinguishes bacterial leafspots from viruses. The lesions often appear in the crown of an orchid or a depression in a leaf where water would usually sit for long periods. If caught early, these bacterial diseases can usually be controlled by removing affected leaves, spraying, and disinfecting pots and benches, with 10 percent household bleach (10 parts bleach and 90 parts water).

Genetic abnormalities can be especially confusing to the grower. This symptom in *Phalaenopsis* is very difficult to distinguish from cucumber mosiac virus (CMV). Often these are seen in flowers that have white areas where there would normally be color, in addition to deformation of the petals and leaves. Usually, genetic abnormalities show no necrosis. You may need to test your orchid to be absolutely sure that genetics and not a virus is the cause of the abnormality.

INDICATOR PLANTS

Indicator plants are those plants which are known to be susceptible to specific viruses. When an indicator plant is inoculated with a particular virus, it exhibits characteristic symptoms. With a little practice, you can learn to easily and routinely answer many questions you might have about the incidence of viruses in your collection by using this method.

Every plant virus has a characteristic *host range*, or a variety of specific plants that are susceptible to that particular virus. When a particular virus infects these host plants, they show typical symptoms of infection and betray the presence of infection in your collection as well.

Although viruses are never good news to the orchid grower, knowledge of plant host range gives you a standard measure of diagnosing viruses for control. To use the method you will *inoculate*, or deliberately infect, indicator plants to characterize the virus. If the inoculated indicator plant shows symptoms of virus, you know the orchid also has the same infection.

Generally speaking, viruses have a wide range of plant hosts. They include tobacco species, bean and pea species, tomato, pepper and spinach, and many common weed hosts as well. For some viruses, host ranges include members of fruit families such as apple and peach.

Symptoms can be *systemic* or *local*. In systemic infections a virus moves within a plant as it grows. In fact, inoculated leaves usually remain symptomless with a systemic infection, but new growth flushes will show the symptoms. These can include mosaic or mottling patterns of light and dark green. In contrast, a local infection reveals itself with symptoms only on inoculated leaves, and does not move systemically within the plant. Local lesions are usually discrete, chlorotic (yellow) or necrotic (dead; red or brown/black) spots on the inoculated leaves.

Which Indicator Should You Use for a Virus?

A great deal of information has been generated for indicator plants to detect CyMV and ORSV. *Cassia occidentalis* (sicklepod) is usually the host of choice for CyMV, but others include

Datura stromonium (Jimson-weed), *Chenopodium amaranticolor*, (dogweed) and *Tetragona expansa* (New Zealand spinach). *Gomphrena globosa* (globe amaranth) is usually the indicator of choice for ORSV. Alternatives include *Cassia occidentalis*, *Chenopodium amaranticolor*, and several tobacco species (see Table 3 on page 46 for summary of indicator plants).

Relatively less information about indicator plants for "other" plant viruses is available.

Since CyMV and ORSV are the major culprits in most collections, you are encouraged to test for them first. If a suspicious orchid tests negative for these two viruses, you can try some other indicators, or contact a diagnostic clinic. The highest risk group is, of course, the large, flowering orchids.

Indicator Plant Seed for CyMV and ORSV

Indicator seed for CyMV and ORSV is sometimes difficult to obtain, but can be purchased from some seed catalogues (See APPENDIX, Suppliers). Generally, most should be grown at temperatures ranging from 65-80 degrees F (18-26C) in medium to low light. These conditions keep the plants tender and succulent. If you plan to use indicator plants, you might devote a small section of a greenhouse to a light shadecloth. That is usually not necessary, though, since most orchid growers tend to keep their greenhouse lighting fairly low anyway, and this should be enough for growth and inoculation of the indicators. It is generally recommended to use one indicator plant per test.

Sow the seed thinly in a light soil mix, about three in each 4-6" pot. *Cassia occidentalis* seed has a thick outer coat, so nick these slightly at the pointed end with a fingernail or wire clipper before planting to hasten germination. *Gomphrena globosa* can be planted the same way, or in a community pot and separated later.

Cassia seedlings can be inoculated as soon as the first pair of three leaflets emerge, although you can also inoculate subsequent leaves as well. Plants of *Cassia* will be ready to be inoculated in two to three weeks after planting. Expect *Gomphrena* to take two weeks to a month longer. *Gomphrena* should have four to six leaves before inoculation. It is especially important that growing temperatures for G. *globosa* do not exceed 90F days/85 F night. Symptoms will not show up as well in higher temperatures (26). Details of two variations on the "rubout" technique are given in Chapter 4.

Indicators for "Other" Orchid Viruses

Unfortunately, less is known about indicator plants for orchid viruses other than CyMV and ORSV. Orchid researchers have discovered that some of the short rhabdoviruses of orchids can be detected using *Chenopodium amaranticolor*, *Chenopodium quinoa*, *Petunia hybrida* and several tobacco species (26). For these viruses, make sure you inoculate under warmer conditions, at 86 degrees F (30C) or higher (9). Chlorotic spots should develop after two to four weeks (9,26).

Bean yellow mosaic (BYMV) which has been reported to infect *Masdevallia*, infects C. *quinoa*, and *Vicia faba* (broadbean) (3, 26).

Tomato ringspot in Cymbidium produces local lesions in *Nicotiana tabacum* cv. Kentucky 35 and *Vigna unguiculata* (cowpea) (26).

Cymbidium ringspot can be detected on *Emilia sagittata* (tassel flower), *N. clevelandii*, *C. amaranticolor* or *Phaseolus vulgaris* (French bean). In each case, chlorotic or necrotic local lesions are formed within four to 10 days (21).

Cymbidium mild mosaic produces local lesions on C. *amaranticolor*. Cucumber mosaic virus (CMV) can be detected using *Citrullus vulgaris* (watermelon), *Pisum sativum* (pea), and *Sesamum indicum* (sesame) (26). Both watermelon and pea can be planted directly into 4 - 6″ pots, with four to six seeds per pot. They can be inoculated as soon as the cotyledons emerge and expand slightly. Sesame should be started as *Gomphrena* would be, in a community pot, transplanting and inoculating at the 4-6 leaf stage.

Rub-out tests using plants for the "other" viruses can be difficult. Seed is sometimes difficult to obtain. In fact, many seed varieties used in rub-out tests are only available through virus research laboratories in universities. In some cases, seed companies do carry certain, less common seed varieties (see

APPENDIX, Suppliers).

It may not be easy to discover or maintain optimum conditions conducive to producing local lesions to detect minor orchid viruses. Luckily, growers rarely encounter "other" orchid viruses. For that reason, rub-out techniques are not really necessary for most growers to try when dealing with viruses other than CyMV and ORSV.

Begin to suspect one of these "other" viruses if an orchid shows suspicious symptoms that persist in new growth flushes and repeatedly test negative for CyMV and ORSV in your normal rub-out tests. In that case, your best bet is to separate the suspicious plant from the others and contact a plant disease clinic in the state or a private lab with adequate diagnostic facilities.

If you are highly motivated, you can try to inoculate a variety of the indicator plants. Keep in mind, however, that negative results are not conclusive evidence that your plants do not harbor one of these viruses. Even if your inoculation does not show a local lesion, the plant could still have an unidentified viral infection.

Rub-out tests for CyMV and ORSV, however, deliver extremely practical and reproducible results. This technique is highly recommended for orchid growers. These two viruses are prevalent in almost all orchid collections, and you as a grower are likely to find some infected plants in tests. Once you purchase seed, you can become your own seed source if you allow plants — preferably from your tested negative controls — to go to seed either in the greenhouse or outside during the normal growing season.

There are other, more obscure orchid viruses this book will not discuss simply because of their extremely limited importance.

TABLE 3. Indicator plants for orchid viruses.

Virus	Indicator plant of choice	No. of leaves to inoculate	Symptoms produced	Seed availability
CyMV	*Cassia occidentalis** (coffee senna)	2 cotyledons	necrotic ll	limited
ORSV	*Gomphrena globosa** (globe amaranth)	4-6 true leaves	necrotic ll	generally available
Orchid rhabdoviruses	*Chenopodium amaranticolor* *Chenopodium quinoa* (pigweed)	4-6 true leaves	chlorotic ll	limited
Tomato ringspot	*Nicotiana tabacum* cv. Kentucky 35 (tobacco)	4-6 leaves	chlorotic or necrotic ll	generally available
	*Vigna unguiculata** 'Black local' (cowpea)	2 cotyledons		
	*Chenopodium quinoa** (pigweed)	4-6 leaves	necrotic ll leaf distortion apical necrosis	
Cymbidium ringspot	*Chenopodium amaranticolor** (pigweed)	4-6 leaves	chlorotic/necrotic ll	limited
	*Phaseolus vulgaris** (French bean)	2 cotyledons	chlorotic/necrotic ll developing into systemic chlorosis and mottle	generally available
	Emilia sagittata (tassle flower)	2-6 leaves	chlorotic ll developing systemic yellow spots and rings	limited
	Nicotiana clevelandii (tobacco)	4-6 leaves	chlorotic/necrotic ll	limited
Cymbidium mild mosaic	*Chenopodium amaranticolor** (pigweed)	4-6 leaves	chlorotic/necrotic ll	limited
Cucumber mosaic	*Citrullus vulgaris** (watermelon)	2 cotyledons	chlorotic/necrotic ll	generally available

*recommended host   ll = local lesion

ELECTRON AND LIGHT MICROSCOPY

Recent scientific techniques have enlarged the options available to test orchids for viruses. Two — electron and light microscopic diagnosis — are accurate diagnostic methods. Unfortunately for most collectors, however, each has limited application for routine virus detection. The biggest limitations are that tests are costly and require training to perform. To a great extent the detection techniques have not moved out of government or university laboratories.

Despite their limitations, knowledge about each method could prove useful to you in the future. Many growers have access to university or state agency labs. Some even offer special diagnostic services for growers in return for specimens. Or, you may someday want to hire a private firm to perform these services for you. Being informed about the latest in testing options will help you ask the right questions and make sure your money is well-spent.

Electron microscopy

Virus particles are small, so small that under an ordinary microscope light waves cannot provide the resolution needed to see individual virus particles. Enter the electron microscope, also called the EM. The EM beam wavelength is much smaller than that of other microscopes. It allows the viewer to actually see virus particles at extremely high magnifications — 100,000 to 200,000 times their actual size.

The EM is a highly sophisticated and elaborate piece of equipment, and its use has been traditionally limited to universities and research centers because the equipment is expensive and requires trained operators.

How to Test a Plant

The *leaf dip procedure* is the most common technique used to view virus particles with the EM. Briefly, a small piece of leaf tissue is minced with a razor blade in a few drops of a buffer solution, such as potassium phosphate, at a near-neutral pH. A drop of this pale green liquid is placed onto a tiny copper grid which has been stained, most often with a heavy metal stain. The stain is called a *negative stain* because the virus particles appear white against a dark background.

The copper grid is then placed into the microscope. The elctron beam passes through the grid and hits a fluorescent screen. The beam is converted to light waves, which the viewer can see (Fig. 9).

A leaf dip takes about five minutes to prepare, but viewing time to search for the virus particles can vary from a few minutes to a half hour or more, depending on the type of virus particles and the *titer*, or concentration, of the virus in the infected plant.

Diagnosing Virus with Leaf Dip

The trained technician can easily discover CyMV and ORSV with the EM. This is because the titer of both viruses in infected orchids is high. A high titer means that the particles can be seen after looking at a only a few grid squares.

Also, many viruses have a distinct "look" under the EM. CyMV particles are flexuous and rod-shaped and about 450 nanometer (nm) long. (With one nanometer equal to a billionth of a meter, or 0.000000039 inches, you can see why the electron microscope is so expensive!) ORSV virus particles are shorter, wider, rigid rods about 300 nm long (Fig. 10). Both bean yellow mosaic virus and vanilla mosaic virus appear as longer, flexuous particles, about 750 nm long. They usually exist in low titers within the plant host so that viewing time may be long in comparison for CyMV or ORSV. This is typical for viruses in the potyvirus family.

Another flexuous, rod-shaped virus is dendrobium vein necrosis (DVNV), which belongs to the closterovirus group. At 1,300 nm long, DVNV particles are the longest of all viruses. The particles are quite distinctive, but they require a special buffered sucrose solution to render them easily visible with reliability (15). Without the solution, negative results with the leaf dip procedure are likely.

Two of the more difficult virus groups to diagnose in

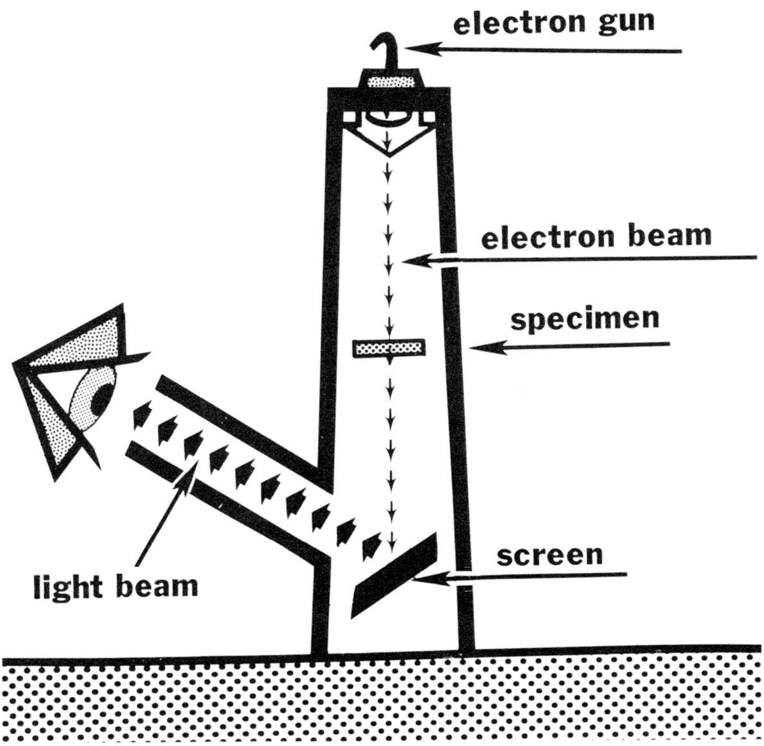

Fig. 9 How virus samples are "seen" with the electron microscope.
(Graphic courtesy F.W. Zettler)

orchids with the EM are members of the rhabdoviruses and the cucumber mosaic virus group. The rhabdovirus group, which has bullet-shaped, or bacilliform virus particles presents a most difficult challenge for EM detection. In contrast, cucumber mosaic, the representative member of the cucumovirus group, is spherical in shape. Without extensive training, a technician can easily overlook or mistake both rhabdoviruses and cucumber mosaic virus for cell organelles of the plant itself.

The Function of EM for the Collector

High cost of the EM, the need for training, and a lack of accessibility to the equipment are major limiting factors of the EM for routine diagnosis of orchid viruses. And, when

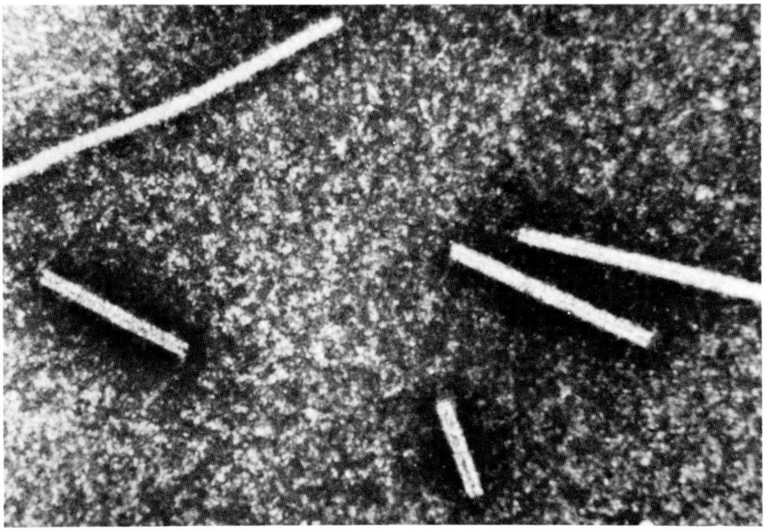

Fig. 10 An electron micrograph reveals the appearance of CyMV and ORSV, both of which infect this plant. The CyMV (1 particle), are longer, slightly flexuous rods, and the ORSV (4 particles), show up as shorter rods.

you consider the large proportion of viruses the typical collection may be infected with, the method is cumbersome and impractical.

EM technology does have its uses, however. A collector may be able to send an isolated sample or two to a university or research center that has an EM and a trained staff to interpret results. It could prove especially valuable to confirm results of other diagnostic techniques or when those other methods, such as plant indicators or serology are not readily available.

Light Microscopy

A light microscope is the type of compound microscope seen in most biology labs. With a light microscope, light is transmitted through the specimen. Viruses can be identified with the light microscope, which works on the principle of staining parts of the virus.

Viruses are made up of proteins and nucleic acids; DNA or RNA. Special stains have been developed which selectively color either proteins (the Azure A stain) or the nucleic acid (an orange/green stain).

In contrast to the electron microscope which magnifies individual virus particles, the light microscope is used to see *virus inclusion bodies*. These are much larger structures which are aggregates of millions of virus particles, forming discrete, characteristic packages. The names of these shapes are described as crystals, plates, amorphous inclusions, pinwheels, scrolls, laminated aggregates, etc. The particular shape of the inclusion body is unique to the virus it reveals.

Researchers at the University of Florida are responsible for developing and using this relatively new technique, one that holds several advantages over the EM. For one, equipment requirements are much less demanding. Light microscopy requires only a good compound light microscope with a 100x oil immersion lens and a low magnification lens of 10x or 20x. Only a few chemicals are needed, the most important being an Azure A stain and a combination of two stains, Calcomine orange and Luxol Brilliant Green (Fig. 11).

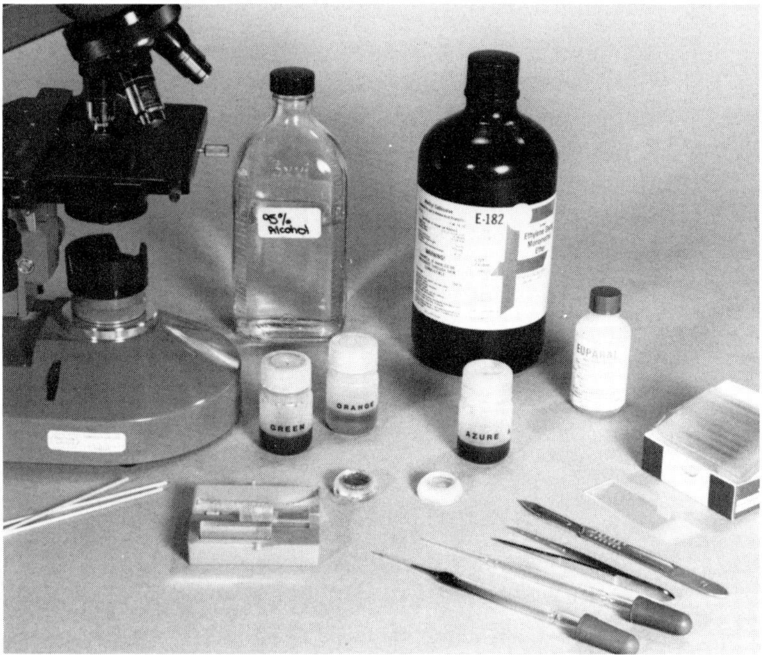

Fig. 11 Materials needed for staining leaf tissue to detect viral inclusion bodies. Pictured here are a light microscope, several stains and solvents, pipettes, microscope slides and small glass dishes, or watch glasses, for mixing stains. (Photograph courtesy of Florida Division of Plant Industry)

Diagnosing Viruses with Light Microscopy

Virologists do not yet know the exact function all inclusion bodies serve, but the fact remains that every plant virus studied so far does form charcteristic inclusions, consistent in whatever host they appear.

Inclusions formed by CyMV, when stained with Azure A appear as purplish, banded, or spindle-shaped bodies. If a green, plate-like inclusion is found in the epidermis tissue of an orchid stained with the orange-green combination, the result is ORSV. In another host, such as tobacco, tomato, pepper or potato, that same inclusion would probably show up as tobacco mosaic virus, (TMV), a relative of ORSV.

Hollow purplish hexagonal crystals in the mesophyll tissues stained with Azure A indicates cucumber mosiac virus or a member of that virus group. Rhabdoviruses typically form inclusions within the nucleus with the Azure A stain. The *Vanilla* potyvirus forms amorphous inclusions, which stain an olive-brown color and are often found next to the nucleus. The dendrobium vein necrosis virus forms striated red-purple bodies in the phloem tissue of infected plants, as is typical of those belonging to their group.

Briefly, thin hand-cut sections are usually just cut with a razor blade or strips of plant epidermis are peeled off with a pair of fine tweezers. Sometimes these sections are treated in various chemicals to eliminate the chlorophyll that will interfere with staining.

Next, the small tissue pieces are immersed in each stain; the Azure A and the orange-green combination for 10-15 minutes. The pieces are then rinsed in 95 percent ethanol, mounted on a microscope slide, and placed on the light microscope for viewing. Viewing time varies. An "easy" sample may take 10 or 15 minutes, while others could require an hour or more. For an excellent and complete review of the techniques and photographs of the inclusion bodies seen in orchids, read, "Light microscopic techniques for detection and diagnosis of orchid virus diseases," by R. G. Christie, N.-J. Ko, and F. W. Zettler, which appeared in the *American Orchid Society Bulletin* in 1986 (Volume 55: pp.996-1007).

The Value of Light Microscopy

This technique has the greatest application to detect viruses for which there may not be antisera available for serological assay, or for those viruses which may not be readily transmitted to indicator plants by rub-inoculation. For example, light microscopy was one of the tests we used to detect the recently described *Vanilla* mosaic virus from French Polynesia and in confirming the latest outbreak of dendrobium vein necrosis in Florida. When we presented a discussion and workshop on this technique at the Mid-American Orchid Congress in Atlanta in 1986, the audience showed a great deal of interest because the virus inside the plant cell could actually be "seen."

The technique probably has most use in confirming suspicions about particular orchids of interest. A warning, however. The method is time-consuming, making it unsuitable for testing every orchid in a large collection.

Light microscopy needs very little special equipment. The technique itself is relatively straightforward, yet the method still remains beyond the technical reach of most commercial growers and hobbyists. Before you could use this method effectively to recognize inclusions and differentiate them from normally occurring cell parts, you would need some training with a person familiar with the method and a lot of practice.

But it is possible to learn, especially if you have some experience in cytological technique and access to a good light microscope. Treatment materials and instructions can be ordered (See APPENDIX, Suppliers).

At the present time, light microscopy is especially valuable for diagnostic clinics or university laboratories that do not have expensive virus detection equipment. It is less valuable for plant virus indexing, or testing programs that require processing large numbers of routine samples in a short time.

Despite limited practical application for large-scale testing of orchids, the technique has led to a greater understanding of many plant viruses and their relationship to viral infection. It has been a valuable aid in understanding the process of virus replication within a plant for optimum purification, as well as for discovering many previously unknown viruses.

SEROLOGY

What is serology? The same principle that makes a vaccine work to protect you against diseases can also be employed to detect viruses in plants. This technique is called *serology*. The technique analyzes the reaction between an *antigen*, such as a virus particle, and an *antiserum*, or antibody that recognizes the antigen. The testing method is based on the fact that an antibody molecule links up to its respective virus particle in an exclusive, "lock-and-key" style that cannot be duplicated with another, unrelated virus.

Although serological techniques are often fairly simple, the production of antisera is somewhat complicated. The virus must first be isolated in pure form. A virus cannot be grown in pure culture as bacteria or fungi can be. It must be extracted from living, virus-infected plants. These purification procedures can be time-consuming and complicated. They involve equipment such as blenders, low and high speed centrifuges, electrophoresis equipment, electron microscopes and spectrophotometers.

Obviously, virus purification and antiserum production are limited to the well-equipped private or research university laboratory. Techniques for its general use are now available with the existence of specific antisera.

Once the virus has been extracted from the plant tissue in pure form, it is then injected in an animal, usually a rabbit, for antiserum production. The animal will not become "infected" with a plant virus but lives a normal, healthy life. Its immune system, however, recognizes the protein as foreign and begins to produce specific antibodies against it. The result is similar to what happens when we are immunized with tetanus, measles or diphtheria.

The antiserum is collected from the rabbit after an appropriate period of time passes to allow enough antibodies to be produced. Only a small amount of blood is taken, usually by nicking the marginal ear vein with a razor blade. The blood is allowed to clot and centrifuged to remove the clot and red blood cells. The remaining plasma, or antisera, contains the desired

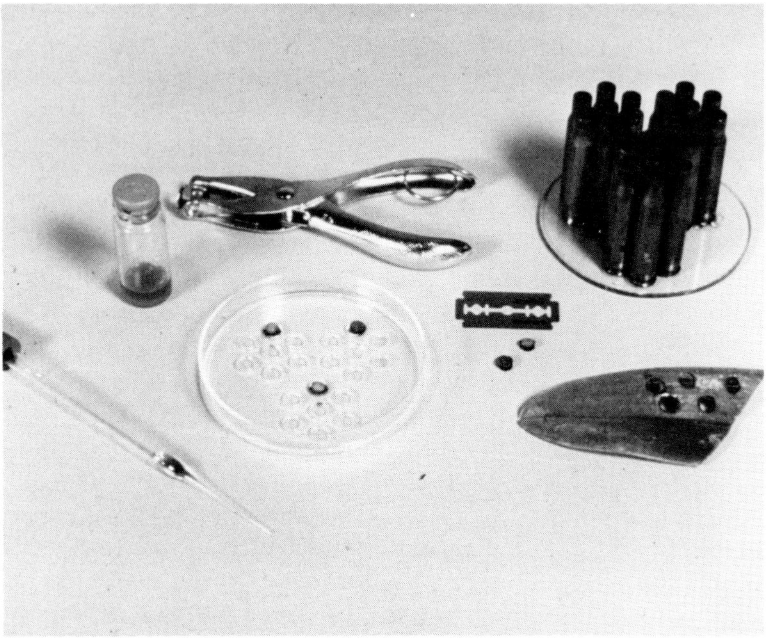

Fig. 12 Materials needed for immunodiffusion assay include pipettes, antisera, agar media, petri plates, paper punches, gel punch and razor blades. The gel punch shown is the homemade-type described in Chapter 4.

antibodies, which are then frozen or freeze-dried in small amounts for later use.

The major shortcoming of serological techniques is that a virus must be purified and a specific antiserum produced before you can detect a particular virus. Ease of virus purification varies greatly among viruses. Some, such as CyMV or ORSV, are relatively easy to purify. On the other hand, good quality antiserum is very difficult to produce for other groups, like the rhabdoviruses. If reliable antisera are not available or cannot be developed for a particular virus, you will have to resort to other diagnostic methods, such as indicator plants, light- or electron microscopy.

Immunodiffusion

This widely used virus diagnosis technique is also known as double radial diffusion (DRD). With DRD both the virus particles and the antibody molecules radially migrate through wells, or depressions, in an agar medium in a petri plate until they meet, link together and form a visible white precipitate. The technique was developed originally to indicate whether a sample of beef was contaminated with another animal species, such as horse or dog. This technique is also used routinely by the Florida Game and Fresh Water Fish Commission to convict hunters of illegal poaching of protected animals like the white-tailed deer and alligator (7). By testing a tiny sample of blood from old clothes or a car seat, officers can tell whether the blood is from a cow, deer, or even an alligator.

Because there seemed to be a demand for a fast and accurate technique which could used in an orchid virus control program, the author modified the DRD technique to safely handle large numbers of orchid samples for CyMV and ORSV (36, 38). The materials needed to conduct a DRD assay are fairly simple and relatively easy to obtain. They consist of a gel punch that you use to punch out a specific well arrangement on an agar medium poured in a petri plate. You can make your own punch or buy one. See Chapter 4 for building instructions and Suppliers for punch vendors. Other supplies you will need for this method are double-edged razor blades, used for sampling leaf tissue, several 1/4″ paper punches, 75 mm petri plates, pipettes and virus-specific antisera (Fig. 12).

Many growers have performed immunodiffusion. Before you try it, though, remember that some training and experience are recommended to differentiate virus-specific reactions from non-specific reactions, which can sometimes occur. If you would like to try immunodiffusion, refer to Chapter 4. Often, there are also opportunities to learn the technique through orchid virus workshops or university classes. Without the skills, it is wise to seek out an experienced virus diagnostician (See APPENDIX, Suppliers).

ELISA, Enzyme-Linked-Immunosorbent Assay

This serological technique is widely used in medicine as well as in many areas of plant virus diagnosis and research. With ELISA, the antibody molecules are linked to an enzyme. When the substrate for that enzyme is added in the last step, the positive, or virus-infected sample changes color to either yellow, green or brown, depending on the type of enzyme used.

The technique is much more complicated than immunodiffusion. Many more steps are required to prepare reagents and carry out the test. Although there are several variations, a commonly used ELISA test is the *double-antibody sandwich* (DAS-ELISA) (12, 13). The test consists of a series of steps that sandwich the antisera with the viral antigen. This part of the test is conducted in a plastic microtiter plate that consists of 96 individual wells. See Chapter 4 for how-to information.

Some commercial and private labs use ELISA for orchid virus testing (1) (See APPENDIX, Suppliers), but many virologists do not consider the test well-suited for routine orchid virus detection tests. This is because a background or non-specific color often appears, confusing test interpretations for those not well-trained in ELISA techniques.

Other Serological Tests

Other serological techniques do exist. Although they are used for plant virus detection, they are not particularly useful for routine orchid virus testing. These include radioimmunoassay, serologically specific electron microscopy (SSEM), and liquid precipitation. As with the other diagnostic techniques, most of the work has been done with CyMV and ORSV.

What is Practical for You

A number of laboratories and universities have developed antisera for purchase. Specific agar media and procedures have been developed for the safe, rapid and accurate diagnosis of CyMV and ORSV by DRD (see APPENDIX, Suppliers). To date, several thousand orchids have been tested commercially by the author with this technique. Interestingly, most clients who have made use of this service have been private hobbyists.

In terms of application to grower needs, the only commercially feasible serological assay for independent growers at the moment is for CyMV and ORSV. Antisera can be obtained for some of the other viruses, such as CMV, BYMV, and TuMV (turnip mosaic), but some experimentation is needed to determine the optimum agar medium and procedures to assay them accurately. Also, since these viruses are far more obscure and less prevalent in the orchid industry, it is not recommended for a routine assay. In some cases, a grower with some experience and training in immunodifussion techniques may prefer to do self-testing with these other viruses, but most are urged to seek professional diagnostic services.

Chapter 3

ACHIEVING COMPLETE VIRUS CONTROL

Precautions and Controls

Too often, orchid growers accept viruses as an unpleasant fact of life that nothing can be done to change. Fortunately, that is not true. You have control over your orchid collection. You are not compelled by nature to have a collection riddled with viruses.

But you must remember that a virus-free collection does not come without some patience and effort on your part. Whether you are starting fresh or evaluating the health of your existing collection, a few simple rules and procedures will help you attain a goal that will assuredly be more than worth the effort you must invest.

Purchasing an Orchid

When buying an orchid, beware of "virus-free" claims. By and large, it is still up to the hobbyist to ask the right questions about the health of the plant. *If* a grower or fellow hobbyist tells you the orchid you are planning to purchase is virus-free, ask first what type of diagnostic test was used to arrive at that conclusion. The knowledge you have gained in the first chapters of this book should help you ask the right questions and understand the diagnostic tests that could have been used.

Ask what viruses have been tested for. Unfortunately, it is not now feasible to test all plants for all viruses, but your plant may have been tested for CyMV and ORSV, the most common orchid viruses.

Ask about the guarantee of the claim. If you see virus symptoms after you buy the orchid, or if your own tests show infection, can you bring it back within a short period of time? Look for a grower who will give you that option, and who will replace your orchid with another — hopefully a virus-free orchid for the one you are returning! If you propagate the orchid, any prior virus infection will be difficult to prove, and the grower may be reluctant to replace it for you. Our experience with several growers has been favorable in this respect. In most cases, a grower is willing to replace an infected orchid. It is good practice to quarantine a new addition to your collection, as Dr. Sproles suggests in the introduction.

Look for a virus-free attitude. When trading or bartering an orchid, be aware of the other person's attitude about viruses. Is the grower blasé about the possibility of a virus infection? Does the conversation reveal that the grower has learned to live with viruses? Attitudes do vary. While some growers may be very serious about maintaining, or working toward, a virus-free collection, others just may not care. After reading this manual, you should be better able to converse easily with most growers on the subject of orchid viruses.

Look at their collection. Better yet, visit the greenhouse and observe growing conditions there. Watch for adequate spacing between plants — they shouldn't be too crowded — and for symptoms of viruses. Ask about the routine for dividing plants, harvesting flowers, insect control, and sterilizing tools and pots.

The Risks of Various Sources of Orchids

Seedlings from community pots These are a good source of healthy material *if* they have not already been divided. If you breed your own orchids, you can feel confident that they are initially virus-free. However, be aware that seedlings are not resistant to viral infection. They can become infected just like any other orchid if you use virus-contaminated cutting tools or if they are visited by virus-bearing aphids.

Wild orchids Orchids collected from their natural habitat are also quite a safe addition to a collection. Again, since neither CyMV nor ORSV appears to be seed-transmitted, you can safely assume that wild orchids are free of these two viruses.

In rare cases, orchids collected in the wild have reportedly been infected with bean yellow mosaic (BYMV), cucumber mosaic (CMV), or any of the insect-transmitted viruses. You are not likely to encounter this problem. Just be aware it does exist.

Also, remember that these orchids, just as seedling hybrids, are not resistant to viral infection once they are placed into cultivation.

Botanicals The small flowering-type orchids, often referred to as *botanicals*, have been under cultivation for a long time, yet they have a low incidence of viral infection. Because botanicals are rarely divided, if ever, and their flowers are rarely cut, CyMV and ORSV are less likely to be found in this orchid group.

Aphids, however, are still a threat and should be controlled as soon as you notice them. Do not let them spread from one plant to another.

Mericlones Meristem propagation in tissue culture is a popular form of vegetative propagation. Unfortunately, many people assume that the simple act of placing an orchid into tissue culture automatically eliminates viruses from an infected plant. In fact, some have suggested that meristem culture has actually done more on a worldwide scale to spread virus by mass propagation than to clean up viruses!

One researcher in the 1950s reportedly eliminated CyMV from infected cymbidiums by meristem culture (29). But a later researcher discovered that the infection in the resulting plantlets was not totally eliminated, even after using the detailed procedure described (5). CyMV is the only virus that has been successfully eliminated from infected plants to a considerable degree through meristem culture (23).

It is wise to remember that meristemming procedures, routinely used in most tissue culture labs, are not generally used for virus control. They are used for mass propagation. That is a distinct difference. To take the extra precautions needed to use meristem culture for virus control, only the actual meristem itself should be excised and propagated on an appropriate tissue culture medium. Unfortunately, the small size of the meristem makes culture and growth of the tissue extremely difficult. To circumvent this difficulty, commercial operations excise a larger piece of the meristem to ensure its survival. Unfortunately, the larger the piece the more likely the virus will be carried along into the cutting and the developing plantlets.

The important point to remember is that the only way that you can assure virus-free plantlets from tissue culture is to test those plants destined for meristeming *first*, only placing into tissue culture those which are healthy. Many careful growers incorporate a virus testing service into their tissue culture program.

Flasked seedlings Viruses can spread through flasked seedlings, even though CyMV and ORSV are not transmitted through seed. Two methods of flasking are commonly used: *green pod* and *dry pod* propagation. In dry pod propagation the pod is allowed to dry and burst naturally, so that the method most resembles natural seedling production. The seeds are then surface-sterilized and *flasked*, or placed in a container of sterile culture medium.

The green pod technique is faster and more commonly used. The immature seed pod is surface-sterilized and sliced open. The placental mother tissue containing the seeds is removed and placed into the culture flask.

The dry pod technique allows the seed to develop and to separate naturally from the mother tissue, so that there is little or no chance of CyMV or ORSV transmission to seedlings. This contrasts with the green pod technique, which transfers the female parent tissue to the culture medium before the seedlings develop. If the female parent is virus-infected, some infected tissue is transferred too. Then, if the seedling and the mother tissue are wounded during the transfer, virus transmission is possible. One study found that one out of 123 *Dendrobium* seedlings derived from the green pod technique was infected with CyMV (42). As you can see, the risk may not be great, but it does exist and you should be aware of the possibility.

One concern that many orchid growers share is the possible

spread of viruses by transferring pollen from a male parent plant to the female. Although the mechanism is not completely understood, neither CyMV nor ORSV are seed-transmitted. Therefore, pollinating a plant with a virus-infected one should be a safe way to produce virus-free offspring. However, when pollinating your orchids, be careful to avoid wounding the recipient mother plant. Sap-to-sap transfer from a virus-infected plant to another — whether you propagate, harvest, or pollinate — can transmit the virus.

Large flowering plants These plants carry the greatest risk of virus infection, simply because they are handled the most. The longer they have been cultivated, the more their flowers are cut or the plants divided the greater the risk of, and the actual incidence of virus infection. This is true, in particular, for CyMV and ORSV. One study by the author tracked the correlation between length of time in cultivation and virus infection. Infection increased from none the first year to five percent in cultivation for one to four years. That figure jumped to 29 percent viral infection for those in cultivation over four years (36).

Some genera with the highest incidence of viral infection by CyMV and ORSV are the *Cattleya* alliance, *Cymbidium*, *Phalaenopsis*, *Dendrobium*, and *Oncidium*. Notice that all produce fairly large, showy flowers.

Establishing Priorities for Testing

Your next step is to establish a priority for testing. If you want to reduce or eliminate CyMV or ORSV from your collection, this step is imperative.

Test all showy, large-flowered orchids first. Most at risk are the showy, large-flowered types, such as members of the *Cattleya* alliance, *Phalaenopsis*, *Cymbidium*, *Oncidium*, etc. Similarly, test first those which are most popular, most frequently divided or propagated, or from which you most often harvest flowers.

Even if you only test these types of orchids, your effort will not be wasted. At the very least you will identify orchids that may be the major source of virus contamination in your collection.

Remember that meristem propagation of these orchid types leaves them just as vulnerable to infection as other less "high tech" propagation methods.

Random sample botanicals and wild orchids. As previously mentioned, botanical orchids are less likely to have become infected with CyMV or ORSV simply because they are propagated less frequently than other types of orchids. Your best option here is to sample selected botanical types at random, noting the virus incidence as you proceed.

Wild orchids can be randomly sampled as well. As long as they have not been propagated, they — as well as seedlings — should not be infected with CyMV or ORSV. Since these orchids can be contaminated by contact with infected tools, pots, media or even hands, remember that they are not resistant to infection.

Once you eliminate any suspicions about virus infection in these groups, you might decide to discontinue testing and concentrate your efforts on the higher risk groups.

Test for CyMV and ORSV first. Since these two viruses are so prevalent throughout the world, you should first emphasize their diagnosis. The presence of CyMV and ORSV can mask the symptoms of other less common viruses. Once you have reduced the incidence of CyMV and ORSV, you may be able to see more clearly evidence of other, less prevalent viruses if they are also in your collection.

Choosing the Appropriate Diagnostic Technique

Assume you have followed the steps outlined above. You have chosen an orchid, gotten some idea of a guarantee and inspected the greenhouse of the grower. You know the inherent risks your choice carries. Now it is time for you to test the orchid yourself.

Which diagnostic method should you use? Those discussed below are ranked by their cost, ease of performance, and accessibility to the average grower. A word of caution: remember that no one diagnostic technique is 100 percent accurate or foolproof.

Indicator plants By far, the least complicated, low cost method of indexing for virus, particularly CyMV and ORSV, is with indicator plants. With some practice you can routinely use this technique for your collection. Practice is important. It is not easy at first to inoculate plants so that the leaves of your indicators are not damaged, which masks the local lesions.

Although the technique is inexpensive and the easiest of any to perform, virus testing by indicator plants does have several disadvantages.

- Availability of the indicator plants. This can be a problem. Many of the plants are weeds and are not available commercially. If, for example, you wanted to use *Cassia occidentalis* for CyMV, you would have to hunt down a supply, probably by writing to several universities or specialized seed companies. This book provides a supplier list of some of the most difficult-to-obtain seeds (see APPENDIX, Suppliers).
- Use of greenhouse space. You will probably also take up valuable greenhouse space with this method, space that could be used for orchids. If you choose this technique, it is a good idea to maintain a batch of young seedlings developing at all times.
- The testing takes time. It may take several weeks to grow some of the indicator plants to the size needed for inoculation. For example, it may take a month or more to grow four to six leaves of G. *globosa*, depending on growing conditions. After inoculation, symptom development may take a week or two more.

The advantages of indicator plant diagnosis far outweigh these inconveniences, however. We advise you to try it on a small scale. Identify some good positive (infected) and negative (healthy) controls in your collections. Think about maintaining your own seed source developed from plants used as a negative control.

If you want to pursue the possibility of expanding an indicator plant test with viruses other than CyMV or ORSV, you may find testing to be a hit-or-miss affair. For example, if a plant tests negative for CyMV and ORSV, but the symptoms still suggest viral infection, you may need more varieties of indicator plants. Luckily, these seeds are usually easy to find in most seed catalogs. Indicators for other viruses include *Petunia hybrida* for some rhabdoviruses, *Vicia faba* (broad bean) for bean yellow mosaic, *Vigna unguiculata* 'Black Local' (cowpea) for tomato ringspot, *Pisum sativum* (pea) or *Citrullus vulgaris* (watermelon) for cucumber mosaic virus.

It is possible that a suspicious orchid can consistently test negative for CyMV and ORSV, yet not respond to routine fertilization and sprays for fungal or bacterial diseases. New growth will still show symptoms. If this happens, you should probably contact a diagnostician for recommendations. It is very likely that your plant has another virus, and the plant should be destroyed.

Electron Microscopy This diagnostic technique is expensive, moderately difficult to perform, and inaccessible to most growers. It is unfortunate that the cost and general unavailability make it inappropriate in most cases for routine assays on any scale because the electron microscope is an excellent diagnostic tool.

Given a little training, a routine "leaf dip" procedure to detect CyMV or ORSV particles or any other rod-shaped virus particle, such as bean yellow mosaic, turnip mosaic, or *Vanilla* mosaic virus is not difficult. Even so, the time involved for preparation and viewing each sample can range from 15 to 30 minutes or more.

If the service is available through a university or diagnosic clinic, you will probably find that the technique is best used with isolated samples. You may consider this technique, however. Be prepared — the costs vary and will probably not be low.

Light Microscopic Detection of Viral Inclusion Bodies This method has wide applications in plant virology, especially to diagnose unknown virus infections and to determine the relationship of viruses and viral-induced protein products. Light microscopic detection is most valuable for detecting some of the less common viruses which do not have indicator plants or other diagnostic means to test, or where routine tests have given suspiciously negative results. The technique costs less than electron microscopy and is generally more accessible.

One of the major benefits of inclusion body detection is that the inclusions are typical of the entire virus group. That means you can assign results to a particular virus group even if the specific virus in that group cannot be indentified. This can give you information about how the virus is transmitted as well as a possible host range.

Another major advantage of this technique is for use in areas where the expensive equipment for serology and electron microscopy is unavailable. If this is your situation, light microscopy may be an essential technique to master.

Unfortunately, it requires a great deal of training and experience to learn to recognize the specific viral inclusions, or aggregates. A grower is not advised to buy a light microscope and the required stains and start making leaf sections. The tests are usually used in universities or plant disease clinics which have the expertise to conduct them. Some may train you to conduct them yourself.

ELISA Again, this serological method is probably not practical unless a laboratory provides the service to you on a fee-per-sample basis. The technique is best suited to established plant pathological/diagnostic labs because it requires a lot of preparation, as well as extensive — and expensive — equipment.

Large numbers of samples can be processed within a couple of days, with results read on an ELISA meter that gives a numerical value related to the amount of color development in each of the wells on the microtiter plate. Test results depend on degrees of color development on the wells, not on simple positive or negative reactions, so it may be difficult to interpret the tests.

ELISA demands positive and negative plant tissue controls, as well as negative controls without plant tissue to determine the background color which often develops. Without these controls, an accurate assessment cannot be made.

The procedure is complex, with the possibility of many potential problems during the tests. Well-trained personnel are very necessary for this test to done accurately. If you are considering the services of a ELISA laboratory, make sure that controls are used and that the results can be reproduced. It might be wise — regardless of what test you do use — to include a particular sample more than once during a submission to the lab so that you can conduct your own quality check on the system.

Double Radial Diffusion Like ELISA, this technique is also widely used in many areas of plant virus diagnosis. The method has been adapted to diagnose CyMV and ORSV with about 95 percent accuracy. Keep in mind that no one diagnostic technique is 100 percent foolproof. It was originally adapted for use by growers themselves for a self-conducted, complete home indexing, or testing, program. DRD has in fact been used by several private hobbyists and commercial growers, some of whom had labs already established for flasking.

This test can be conducted at home even without a lab, however, using ordinary kitchen equipment, such as a pressure cooker, stove and plastic sealable containers. But even with the basic simplicity of the tests, some training is needed, not only for procedures but for correct interpretation of results. For this reason a grower without experience in the tests should seek a reliable, established testing service.

As with ELISA, be sure that appropriate controls are always used and that personnel are well trained. In most cases, the cost of the test per sample is quite reasonable, especially considering the average cost of an orchid plant.

Sanitation

Good sanitation habits in your greenhouse are essential! If you do not use good sanitation procedures in your greenhouse you will not be able to control manually transmitted viruses, the major viral troublemakers for most hobbyists and commercial growers. Generally speaking, the more stable viruses that survive well on tools and pots are the least likely to have an insect vector. Conversely, those viruses that do not survive well on cutting tools tend to have an insect vector.

There are basically two ways to inactivate a virus on contaminated tools, wire hangers, and pots: heat, or by oxidation with alkaline chemicals (25, 26).

Heat inactivation You can use your home oven to decontaminate tools and pots. First, however, quickly rinse them with a forceful spray of water to make sure the pots and tools are free of plant residue. Be sure to check the shear hinges for plant debris, too. Place the cleaned tools and pots in an oven set at 300 degrees Fahrenheit, for one hour.

An alternate method is to treat the cleaned knives and shears in a 70 or 95 percent ethyl alcohol dip. After you dip the tools, heat them to red hot with a propane torch.

For light cutting and flower harvesting use either disposable single-edged blades, which can be heat-treated, or double-edged razor blades that should be snapped into quarters while still in their wrapping. Both are inexpensive and convenient, but remember to discard them in a special "sharp" container for your safety.

Chemical inactivation You can also deactivate viruses with a strong alkali solution. A 10 percent (10 parts plus 90 parts water) strength household bleach solution is often used for dipping tools and soaking clay and plastic pots. Unfortunately, a simple dip by itself is not enough to inactivate all viruses.

First rinse tools with a strong water spray, making sure that all plant debris is removed. Clay and plastic pots should also be scrubbed to completely remove any roots or media that might adhere to the pot.

Soak the tools for several seconds. Soak the pots for at least an hour. If you are cutting many plants at a time, it is a good idea to own a few sets of cutting tools. That way, you can make sure you let the tools soak for as long as they should. One word of caution: the bleach is corrosive to metal, so wash tools thoroughly at the end of each day you work.

Because bleach is toxic to some plant tissue, be sure to rinse the tools with water after soaking. Pots should also be completely free of bleach. You can accomplish this best by rinsing them with water *twice* to remove any bleach residue.

As the bleach begins to turn green from plant sap, it loses its virucidal activity — another good reason to rinse your tools before soaking. You should change the bleach as soon as the solution begins to turn green.

Avoid exposing the bleach to light when you use or store it.

Store in a dark bottle or a solid-color plastic bottle. Before you use a fresh batch of bleach, conduct an informal "nose test." A good "bleach" smell tells you it is still an active virus-killer.

Another good dip for tools and pots is a saturated solution of trisodium phosphate (TSP). You can buy TSP at any hardware store, and in some drug stores as well. To make a saturated solution, slowly stir the TSP into a certain volume of water (a clear gallon jug is convenient), with or without a very low heat, allowing it to dissolve. After enough TSP is added, the water can no longer accept the salt, and crystals will begin to fall out of the solution.

Stop adding TSP at this point. Stir it, and/or just let it sit overnight. Crystals that never dissolve indicate that the solution is now saturated. The high pH, like a bleach solution, will inactivate the virus particles.

Soak tools and pots for the same amount of time in TSP as you would bleach. The same precautions hold for TSP as for bleach: rinse tools before and after a dip; remove adhering plant debris, and discard the solution as it begins to turn green. TSP, unlike bleach, is not corrosive to metal.

Here are some other sanitation tasks that you must perform to maintain a high quality, virus-free orchid collection.

1. Always use new, sterile media when repotting plants.

2. Remember that your hands can spread contamination. Virus particles are very small and can survive underneath your fingernails or other cracks and crevices in your hands.

Your hands cannot withstand the abuse of a soak in bleach or TSP. Instead, you might try wearing plastic or rubber gloves that are smooth, free of crevices and can be dipped and rinsed easily and discarded afterwards. Dip them for a few seconds after you cut each flower or divide a plant. Change them often to disinfect, or discard.

3. Spray down greenhouse benches with a 10 percent bleach or saturate TSP solution.

4. Scrub or scrape the bench to remove any

roots and plant debris which may have adhered to it. Remember to treat your brushes as well.

5. Space plants in your greenhouse, avoiding overcrowding. The tough leaves of most orchids make it unlikely that injury will occur simply by the orchids touching, but remember that insects can crawl from one plant to another if your plants are too crowded.

6. Arrange plants so that those hanging from the wall or ceiling, or perched on a higher bench, do not drip on lower plants. This is especially important for high risk orchids.

7. Destroy virus-infected plants, preferably by burning. If they still bloom well and you cannot stand the thought of destroying the plant, isolate it on a separate bench or in a separate greenhouse apart from both high risk and healthy orchids.

8. Always work on the known healthy orchids first and then move to work on the infected ones. Even better: work on virus-infected and healthy plants on different days.

9. Yes, smoking tobacco can carry virus. TMV, a close relative to ORSV, exists in smoking tobacco depite the curing process. TMV from tobacco has been rubbed out on orchids and causes only a local lesion reaction similar to that seen in indicator plants. TMV does not move systemically in orchids. However, this virus does have a wide host range and can easily infect many other plants, including members of the Gesneriaceae, which are covered in the Appendix. Do not smoke in the greenhouse. If you have smoked before entering the greenhouse, wash your hands thoroughly before handling the orchids.

10. Maintain good air circulation with fans and avoid overwatering, especially during the cooler seasons when certain fungi and bacteria

are prevalent. It is best to water early in the day so that standing water will evaporate.

Do these precautions sound like a lot of extra work? If they do and you are not on this sanitation regime now, you are probably also spending extra time and effort dealing with viruses — not to mention the loss of vigor and plant quality caused by other diseases as well as the viruses. When weighing the advantages of a rigorous sanitation schedule against the extra time and effort required, remember to factor in the extra benefits you will receive when you reduce other fungal and bacterial pathogens that attack a virus-weakened plant.

Also, remember that good sanitation practices are a habit. Like any other habit you must learn, sanitation will eventually become a natural and almost effortless part of your orchid management routine.

In short, sanitation pays off!

Insect Control

Aphids Aphids are the most notorious insect vector for viruses. These pear-shaped insects are from one to six millimeters long — less than 1/4 of a inch — yet they account for over half of all insect-vectored plant viruses. They also account for more than 71 percent of all vectors in their insect family of Homoptera, which includes aphids, scales and mealybugs (46).

The piercing and sucking mouthparts of aphids are designed for transmitting viruses. When they pierce a leaf and suck the juices, they excrete saliva which in itself has a toxic effect on a plant. The salivary secretions also contain viruses which are transmitted from plant to plant.

Species range in color from white to green, yellow or black. The insects reproduce rapidly with the help of parthenogenic females that give birth to live young without even mating. When colonies on one plant become crowded, aphids can develop wings and fly away to establish new ones.

Aphids are often found in orchid collections, particularly on young buds and inflorescences, and any colony should be regarded as suspect and unwanted. Common as they are to orchids, however, they will seldom form extensive colonies on the plants (19, 46), simply by preference. Most common species on orchids that act as vectors are *Aphis gossypii*, the cotton-melon aphid; and *Myzus persicae*, the green peach aphid.

Table 2 lists several reported orchid viruses, the probable group to which they belong, and the vector associated with that virus, as previously mentioned. Virus transmission by aphids has been experimentally demonstrated with bean yellow mosaic, clover yellow mosaic, cucumber mosaic, dendrobium mosiac, turnip mosaic and the *Vanilla* potyvirus (46). Based on knowledge of related viruses, aphids have been accused of virus transmission for dendrobium vein necrosis and the rhabdoviruses.

An aphid species and its associated virus have evolved a highly specific relationship. The insects can transmit specific viruses in three different ways. Knowing about how they transmit the viruses can help you control them.

Non-persistent transmission of a virus occurs when an aphid feeds on a plant. The virus particles adhere to the insect's mouthparts, called stylets, and are injected into a plant on the next feeding probe a few seconds or minutes later. With this type of transmission, the virus is merely passed on by the insect. It does not remain in the aphid after feeding.

Persistent transmission of a virus occurs when an aphid actually injests the virus. The virus particles move all the way to the midgut, entering the insect's blood system over a period of time. The particles finally enter the salivary glands to be injected back into a plant during a later feeding. It may take several days for the aphid to transmit a virus in this way, but the insect carries the virus — and can transmit it — throughout its lifetime (46).

Semi-persistent transmission occurs when a virus accumulates in the aphid's foregut and is injected into a plant during feeding. This type of transmission may take several hours and is intermediate in terms of the length of time required to transmit the virus between non-persistent and persistent viruses.

Applying biology to your advantage How does the mode of transmission relate to virus control of insect-transmitted viruses in orchids? Viruses within the same families tend to be transmitted over similar lengths of time. For example, cucumber mosaic, bean yellow mosaic, clover yellow mosaic, dendrobium mosaic, turnip mosaic, and vanilla mosaic and all other potyviruses are transmitted non-persistently. This means that the aphid can transmit the virus immmediately after acquiring it. In contrast, dendrobium vein necrosis is a member of the closterovirus group. It is probably transmitted semi-persistently. This means that the aphid transmits it several hours after acquiring the virus.

Rhabdoviruses are usually transmitted persistently, so it may be several days before the aphid can transmit these viruses.

You can apply this biological information to your insecticide spray program. First, be on the lookout for aphid colonies. Spray as soon as you detect any aphids. At the same time, isolate the plant so that the aphids cannot crawl from one plant to another. Although aphids do not typically colonize orchids, remember that it only takes a few aphids to transmit a virus.

If you control the outbreak quickly with the appropriate spray, the winged forms should not develop at all. Because isolation makes it more difficult for insects to move between plants, aphids will most likely die before they are able to transmit the viruses to another plant. Catching colonies before they have a chance to become established and using the correct spray to eliminate colonies is the best way to handle the possibility of an outbreak of an aphid-transmitted virus. Again, due to frequent changes in pesticide laws, it is wise to consult an agricultural agent or local nursery for updated recommendations.

Other insect vectors When the subject is controlling viruses caused by insects, aphids justifiably get the lion's share of attention. Although other insects such as mealybugs, beetles and thrips also think that orchids are tasty, aphids are the most likely viral threat.

Almost, that is. An unlikely cause of possible transmission also exists with nematodes belonging to the order *Dorylaimida* (20) and tobacco rattle virus and tomato ringspot, both of which are reported to infect orchids (26,46). Nematodes are rarely a problem in orchid collections, however, because of the widespread use of sterile potting media. Terrestrial orchids grown in the absence of sterile potting media can, however, become infected this way.

Other greenhouse pests Growers are often concerned about the possibility of snails, slugs, ants, roaches, or even mice transmitting viruses. None of these pests has ever been proven as virus vectors of orchids or any other plant.

Handling pesticides

Insect control is an important part of maintaining a healthy collection. Remember, however, that you must handle and apply pesticides in strict accordance with directions on the pesticide control label.

One insecticide which may be tried experimentally for control with safety is Safer's Insecticidal Soap (See APPENDIX, Suppliers), which is mixed at 2.5 ounces or 5 tablespoons (75 mls) per gallon. Whatever you do use on your orchids for insect control, you are encouraged to try a small test area on a plant before spraying an entire collection to be sure that there will be no toxic effect.

Here are some general rules to follow when using any pesticide (33):

1. Follow the recommendations on the label. Although this rule sounds obvious, many people don't bother to follow it.
2. Spray at a distance of 18 to 20 inches from the plant to avoid problems with toxicity.
3. Wear protective clothing when mixing and applying pesticides. These include rubber gloves, spray suits, goggles and a vapor respirator.
4. Mix pesticides in a well-ventilated area. Avoid contact with eyes and skin.
5. Store pesticides safely away from extreme heat or cold in a place where children and animals have no access.
6. Do not smoke, eat or drink while mixing or applying pesticides.
7. Do not reuse pesticide containers. Crush or puncture containers to avoid reuse and dis-

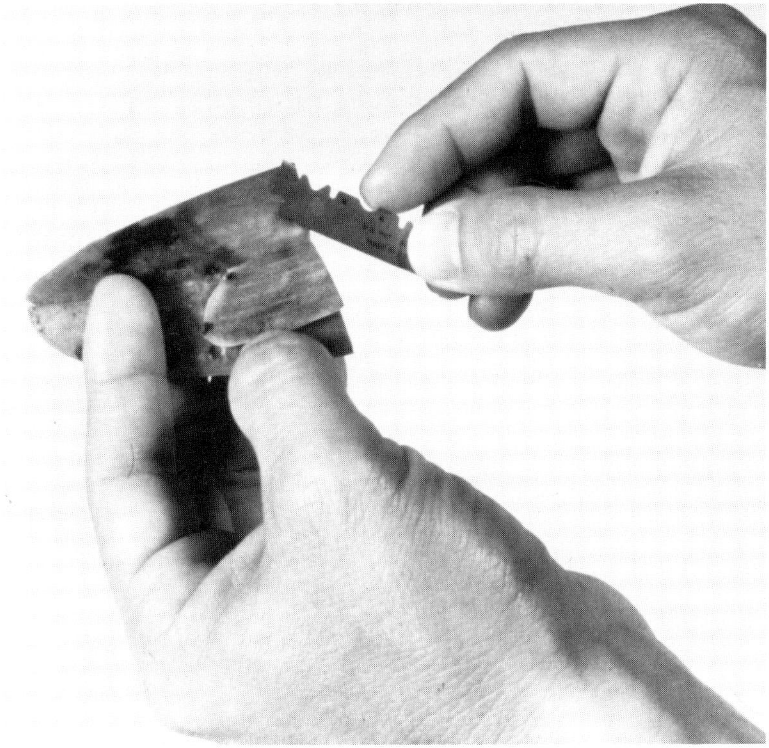

Fig. 13 Slicing a leaf section to expose plant sap containing the virus particles.

card in a garbage can for disposal in a sanitary landfill.

Growers often ask about virucides which can be sprayed onto an infected plant to cure of a viral infection, just as you would spray for a fungal or bacterial disease. Some research has been conducted with a virucide called Virazole (31) in relation to suppressing human viruses. Little research has been conducted with plant viruses, however (26). It is unlikely that effective virucides will be available soon, but may in the future. Genetic engineering is responsible for some exciting new developments in the plant kingdom. As research progresses, growers may be able to buy genetically engineered orchids that have resistance to viruses built into their genetic makeup.

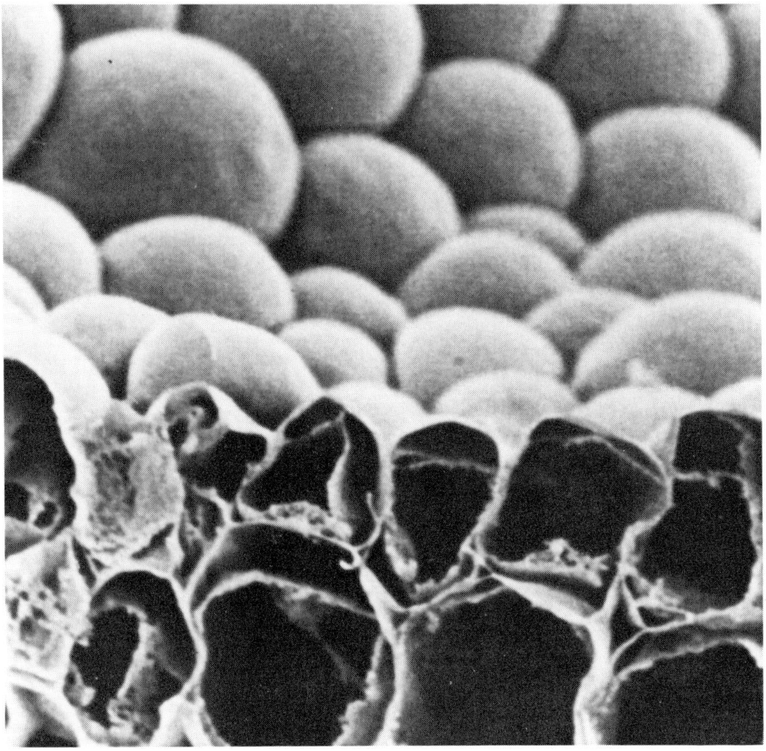

Fig. 14 A photograph from a scanning electron microscope showing cell eruption by the slicing of an orchid leaf.

(Photograph courtesy Mark Whitten)

Chapter 4

VIRUS IDENTIFICATION: STEP-BY-STEP

INOCULATING INDICATOR PLANTS

How to Inoculate: Two Methods of the Rub-Out Technique

You can use either of two simple inoculation techniques. The first is easier and requires only a clean razor blade.

The Razor Blade Technique

This technique (24) is most convenient for thick-leaved orchids, such as members of the *Cattleya* alliance, *Phalaenopsis*, *Oncidium*, etc.

1. With a clean razor blade, slice a piece of tissue from the orchid plant to be tested. A three- or four-inch piece is usually enough.
2. Slice the leaf sample of the orchid to be tested just parallel to the leaf surface so that about 1/2″- to 1″- square surface of cut leaf cells is exposed (Figs. 13, 14). Preferably, the selected sample should be taken from the previous year's growth.
3. Lightly dust the leaf surfaces of the indicator plant with a 500-600 mesh Carborundum (See APPENDIX Suppliers). You

Fig. 15 Rubbing the exposed cells onto carborundum-dusted indicator plant leaves (*Cassia occidentalis*).

should be just barely able to see the blueish color of the Carborundum on the leaves. The abrasion from the Carborundum produces microscopic wounds on the leaf surface, allowing the virus particles to enter cells of the inoculated leaves. The Carborundum can be used in both rub-out techniques. It may not be absolutely necessary as wounding occurs just by rubbing the hairs on the leaf surface, but it does help and is usually recommended.

4. As soon as you cut the orchid leaf, rub it gently but firmly onto the leaf surfaces of the dusted indicator plant (Fig. 15). Try to rub as much of the leaf surface as possible. Take care that the rubbing does not damage the plant's leaves. A little practice will help you get the right feel for this technique. You can tell if you were a bit overenthusiastic if the leaf shows obvious signs of scratching or turns brown and dies one to two days after the inoculation.

5. After rubbing, gently wash or mist the inoculated leaves with water.

6. Place on an isolated bench in the greenhouse and observe for symptom development for one to two weeks.

For thin-leaved orchids like *Phaius*, a variation of this technique works well.

1. Rub a clean piece of sandpaper on the orchid leaf to be tested, wounding and scratching the surface so that the plant sap is released.

2. Then rub the wounded leaf over the surface of the indicator plant as described above.

3. After rubbing, gently wash or mist the inoculated leaves with water and place in the greenhouse.

Grinding Tissue to be Inoculated

This procedure is most often used in virus testing labs.

1. Sterilize a mortar and pestle by soaking in a 10 percent household bleach solution for 30 minutes. Rinse thoroughly and bake in a 300 F oven for one hour.

2. Select a sample from the orchid to be tested. Cut with a clean razor blade and place into the sterile mortar with a few drops of a buffer such as 0.5 percent (0.5 parts in 99.5

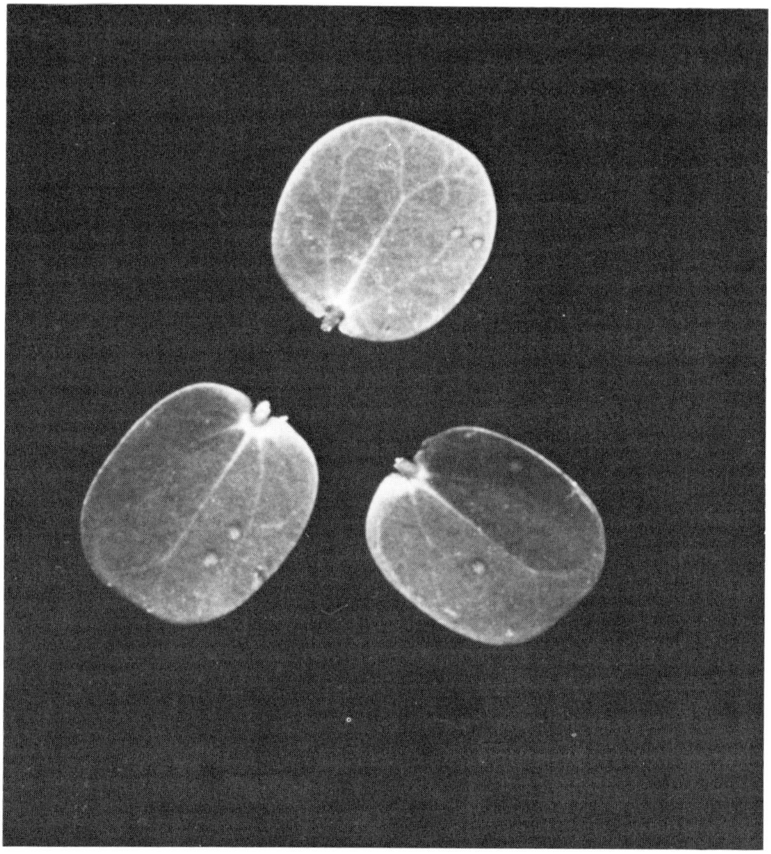

Fig. 16 These pinpoint, red-brown local lesions on *Cassia* are caused by inoculation with CyMV.

parts) potassium phosphate (K_2HPO_4), or plain water.

3. Lightly dust the indicator plant with Carborundum.

4. Crush the leaf cells with a sterile pestle. This releases the virus particles. Grind until the tissue is well crushed to make a slurry. It may help to dice the leaf with a clean razor blade first.

5. Working quickly with a cotton swab, rub the sap onto the leaves of the dusted indicator plant.

6. Gently rinse or mist the leaves with water and place in the greenhouse for 1-2 weeks to observe symptom development.

Determining Results

For both CyMV in *Cassia* (Fig. 16) and ORSV in *Gomphrena* (Fig. 17), discrete necrotic lesions will appear if the test plant was infected. Reaction can take from a few days to two weeks. Observe and compare these reactions, recording results.

Negative and positive controls Whenever you inoculate you should also run negative and positive controls. A young seedling which has never been divided or an orchid recently collected from the wild should work well as your negative control. The negative tests are a check on your indicator plant production environment.

A *negative control* like this is also an indication of environmental factors that can cause damage to indicator plants. For example, in areas with high levels of air pollution, you may observe nonspecific, pinpoint necrotic lesions on *Cassia* leaves (26) that have nothing to do with virus. Damage from pesticides could also be confusing.

Positive controls are also important. It is wise to test your most suspicious plants first, especially those that have shown virus symptoms like flower necrosis or color break. Ask other growers if they might sample one of their suspicious plants for testing. This will allow you to verify that your conditions are accurate.

Naturally, once you've mastered the techniques and find virus-infected orchids, you'll maintain those plants well away from other orchids. But use some of these plants every time you run a test, too, just to make sure that your techniques and environmental conditions are adequate for local lesion assay.

Without both positive (infected) and negative (healthy) controls, you will not be able to evaluate your results realistically. In the short run, positive and negative control tests take a

Fig. 17 Necrotic, reddish, local lesions on this inoculated *Gomphrena globosa* are caused by ORSV. (Photograph courtesy R. H. Lawson)

little extra time, but they will give you more confidence about your results in the long run.

THE DOUBLE-ANTIBODY SANDWICH (DAS) ELISA TECHNIQUE

The ELISA technique has many variations. A general procedure is given in this section. Briefly, during the process virus-specific antisera are first loaded into the microtiter plate. The plant samples are added next. Then, an enzyme which is linked, or *conjugated*, to a purified antibody is added. After this step the antibody-enzyme conjugate binds to the virus like a sandwich if a virus is present, with the virus in between. If no virus is present, the antibody-enzyme conjugate will be removed when the plate is washed. The final step is the addition of a substrate which reacts with the enzyme, if present, giving a color reaction. This color can be yellow, green, or brown, depending on the enzyme used.

Equipment and supplies needed:

refrigerator
sink
mortar and pestle
microtiter plate
cheesecloth
cellophane
ELISA meter
purified antisera
sodium carbonate coating buffer
phosphate buffer
Tween-20 (PBS-t)
p-nitrophenylphosphate (substrate)
diethanolamine buffer (substrate buffer)
alkaline phosphatase (enzyme)

Antisera used with the ELISA test are not generally applied in a crude state as they are for immunodiffusion. They are

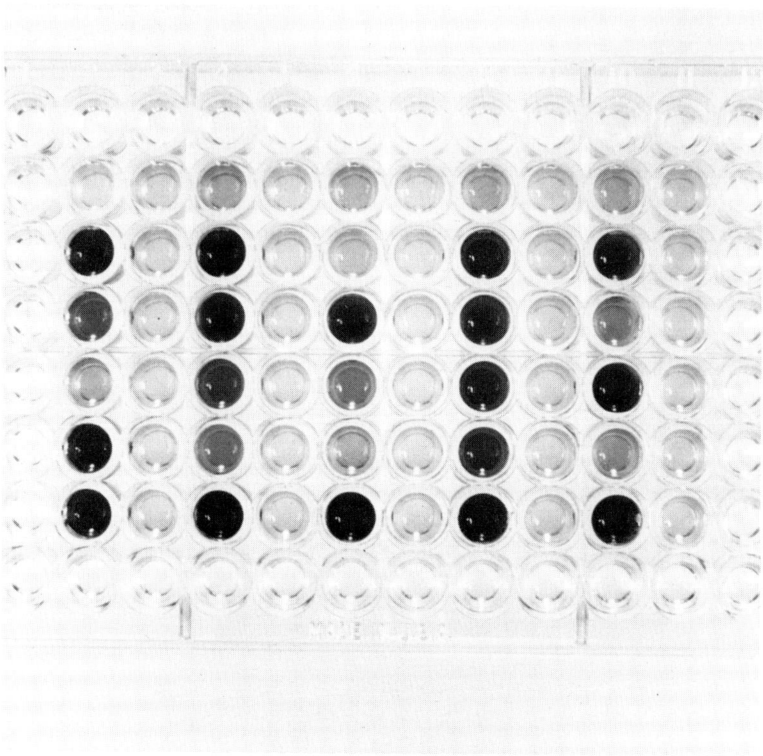

Fig. 18 Reactions of an ELISA test, showing differing intensities of color.

usually purified free from other serum components in a process that can take several days. After the purified antibody and the enzyme-conjugate is made, or you have purchased some purified antisera, here are the steps you should follow:

A microtiter plate with 96-100 individual wells is loaded with the specific purified antibody (0.05-0.2 ml per well) at an appropriate dilution of 1-.001 percent, in a sodium carbonate coating buffer, with a pH of 9.6. The best dilution is determined in a trial test. Plates are wrapped in cellophane or placed in an air-tight container and incubated at 4 degrees C (38F) (refrigerator) overnight, or at 37 degrees C (99F) for one hour. In this step, the specific antibody binds to the surface of the microtiter wells.

After each step, plates are washed three times in a phosphate buffer solution, with a pH of 7.4. The solution contains 0.05 percent Tween-20, a wetting agent (PBS-t). In this step, the antibody remains bound to the plate.

The test samples are added to the wells of the microtiter plate. Samples are usually ground in a mortar and pestle with PBS-t at an appropriate dilution. One part tissue to 10 parts PBS-t is common, but this varies with the test. Again, the plates are wrapped, incubated, and washed. If the virus is present in the plant sample, it will bind to the specific antibody already bound to the plate. The washing will remove extraneous plant debris.

The enzyme-conjugated antibody is added next at an appropriate dilution. One part conjugate to 1,000 parts buffer (PBS-t) is commonly used. If the virus has bound to the antibody added in the first step, the antibody-enzyme conjugate now binds to the virus like a sandwich, with the virus in the middle. If no virus is present, the antibody-enzyme conjugate will be removed when the plate is washed.

A substrate solution, p-nitrophenylphosphate, is the final step. This solution gives a yellow color reaction when it comes in contact with the enzyme alkaline phosphatase. This substrate comes in tablet form and is mixed at one tablet to 5 ml of substrate buffer (diethanolamine in water). Samples testing positive for the virus will begin to turn yellow within a few minutes after adding the substrate. The negative, or healthy samples, will remain colorless (Fig. 18). Visual results are recorded by using a + or − notation, or with an ELISA meter which measures the amount of color at a wave length, in this case 405 nm. The ELISA meter gives a numerical value to the amount of color produced.

IMMUNODIFFUSION

The Ouchterlony agar double radial diffusion technique was chosen for a grower-oriented virus testing technique because of its speed and accuracy, and especially because many orchid growers are already familiar with agar gel systems used in seed and meristem culture. The basic tools to perform this test are simple and inexpensive.

Equipment and supplies needed:

refrigerator
pressure cooker
vacuum pump or faucet-type respirator *
source of a flame (alcohol lamp, candle)
gel punch
6mm paper punch (several)
plastic petri dishes
tweezers
pipettes
plastic microtiter plates
sterile forceps
paper towels
black box *
log sheets to record results
antisera
agar media:
- Noble agar (0.8%)
- sodium dodecyl sulfate (SDS) (0.5%)
- sodium chloride (1%)
- Trizma base (0.6%)

epoxy resin
distilled or ionized water
ice
boiling water
alcohol

*Optional

The double radial diffusion test involves a specific reaction between an antigen (virus particle) and an antibody molecule which recognizes that particular virus whether it be CyMV or ORSV. Antisera which contain these respective antibodies are produced by immunizing (injecting) animals, usually rabbits or goats, with purified virus particles. Small amounts of blood are then taken from these animals weekly. The red blood cells are removed by centrifugation. The remaining plasma, or antisera, containing the specific antibodies is stored frozen until used for testing. This antisera can also be freeze-dried, stored indefinitely in the freezer, and resuspended when needed simply by adding water.

Steps for conducting this test are:

1. The immunodiffusion medium used in orchid virus assays consists of 0.8 percent Noble agar, 0.5 percent sodium dodecyl sulfate (SDS), 1 percent sodium chloride, and 0.6 percent Trizma base.

Add the ingredients all at once, adjust the pH to 8.0, place in an autoclave or pressure cooker, and bring up to 15 psi for 5 minutes until the agar is melted.

Cool the solution to lukewarm. Pour into plastic petri dishes, 12 ml per plate. After the agar has solidified, refrigerate the plates until very firm. When cold, the media will become opaque-white.

2. Cut wells with a gel punch made from 21 7mm Mauser cartridges arranged with a central well and six peripheral wells spaced 5mm from

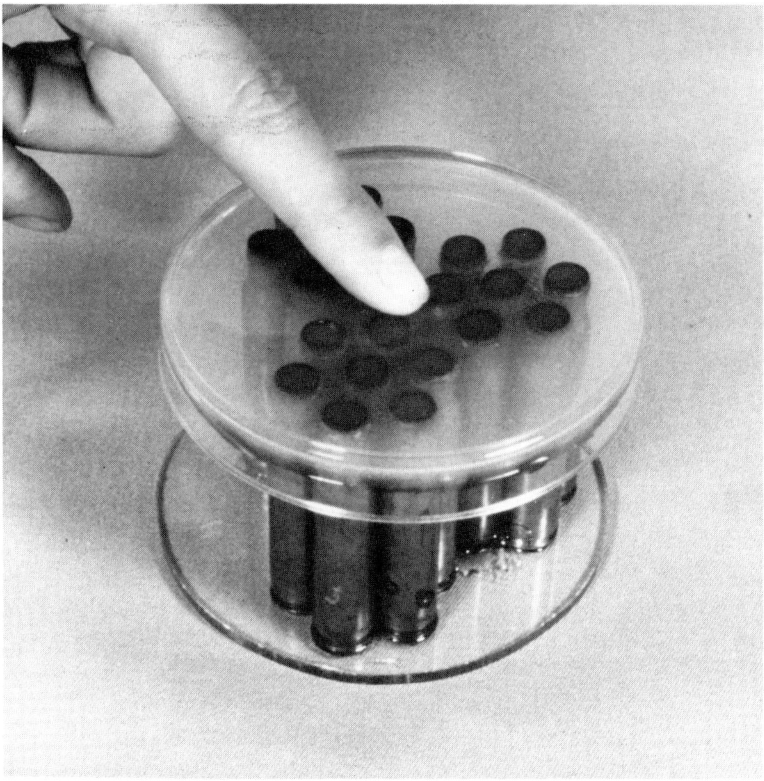

Fig. 19 Homemade gel punch is used to punch out the well pattern in the petri plate of agar medium.

each other. Attach these to a small petri plate which has been almost filled with epoxy-resin. Three of these seven-well patterns are usually cut in each plate (Fig. 19). The agar discs are then sucked out by a small glass tube which is attached by rubber tubing to a vacuum pump or a faucet-type aspirator (Fig. 20). Store plates in the refrigerator. Agar discs can also be picked out with tweezers.

3. Before setting up the test, thaw the antiserum and place in an ice bath. If you use freeze-dried antiserum, add 0.9ml of distilled or deionized water per ml of original volume. Swirl

until completely resuspended. The antiserum should be stored frozen whether resuspended or not and should not be warmed to room temperature.

4. With a pipette, place one drop of distilled or deionized water in each of the 18 peripheral wells of every plate. This ensures contact between the disc and the agar medium. Central wells receive the antisera after plates are loaded with the plant samples.

5. Test samples consist of leaf discs made with a metal 1/4″ (6mm) paper punch which is dipped in alcohol and flamed between each sample. Make two discs for each plant sample, one for each virus test. For thick leaves, such as *Cattleya*, one disc per agar well is sufficient. However, thin leaves, such as *Cymbidium*, *Phaius*, etc. will require two to three discs per agar well, enough so that the discs come to the top of the well, fairly even with the agar surface. Always include a "known" infected sample and a healthy sample as controls.

For CyMV fresh leaf discs can be added directly to the peripheral wells in the order that samples are recorded onto log sheets (Fig. 21). However, you must heat discs for ORSV prior to testing for optimum sensitivity. Plastic microtiter plates, available through most scientific catalogues, with wells the size of the discs, are convenient for this purpose.

Place discs to be tested for ORSV in the wells of the microtiter plate in the same order as those for CyMV. These are suspended about one inch above a boiling water bath for five minutes. These heated discs are then placed in the agar wells of the plates designated for ORSV, in the same order as for CyMV. For each one of the 18 samples you should have two plates, one for each virus. Be sure and use sterile

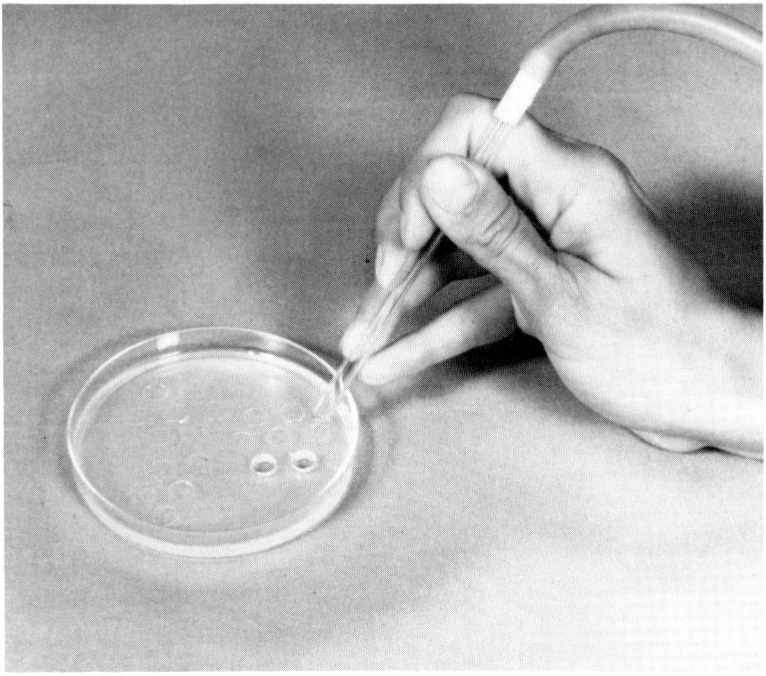

Fig. 20 An aspirator is used to suck out the agar discs.

forceps to handle the discs. Dip them in alcohol and flame between each sample.

Almost as soon as the leaf discs are added, the agar will absorb pigments and turn dark. This will not interfere with the reading of results.

6. After you add the leaf discs, take a pipette and place two to three drops of antiserum into the central wells of each plate (Fig. 22). The serum should be concave with the agar and should not overflow. Bubbles in the antisera can be eliminated by stabbing them with the twisted end of a paper towel. While working with these plates, avoid scratching the bottoms on the working surface. Scratches will interfere with the reading of results.

After both the leaf discs and the antisera have been added, check each plate to see if each disc is sitting in a "cushion" of water. Add more water if needed.

7. Place the loaded plates in a closed container, such as a Tupperware container, with wet paper towels lining the bottom to keep plates from drying out overnight. Store at room temperature. Using this technique, two people can process about 100 samples per hour, with results delivered within 24 to 48 hours. A white precipitin line should form between the center wells and those plant samples which were infected with the respective virus (Figs. 23, 24).

Both the virus particles and the antibody molecules diffuse radially through the agar medium. When the respective particles meet and build in numbers, they form a "lattice" which is seen as a thin white line. At times these lines may be faint, so you need to look at the plates carefully.

8. In order to read and record the results, place the petri plates above a dark background with diffuse light coming in from the side. Alternatively, you can construct a black box which is excellent for reading reactions. To build one, attach a circular light to the bottom of a shallow cardboard or wood box which you have painted black on both the inside and outside. Line the center of the circular light with black felt or velvet to eliminate glare. Cut a hole the exact size of the petri plate at the top of the box to view reactions. Reactions can then be drawn on the log sheets. A log sheet example has been provided for you to copy on page 100.

The technique of virus detection described is reliable and relatively simple to perform. To date, several commercial and private orchid growers have successfully performed this test for CyMV and ORSV after some training.

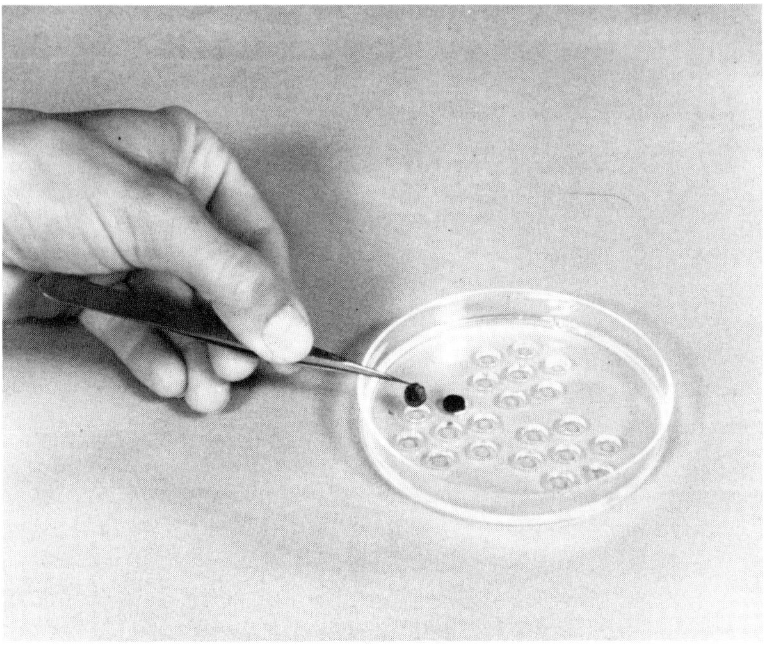

Fig. 21 Leaf discs are placed into the 18 peripheral wells on each plate.

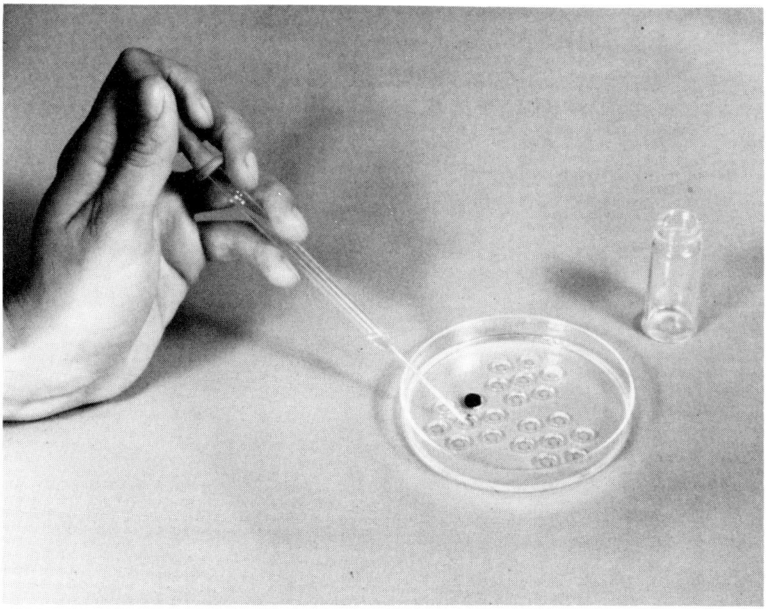

Fig. 22 Antisera is added to the central well of each of three groups of six plant samples. In this photo only one leaf disc has been added. Usually, it is best to add all plant samples before adding antisera.

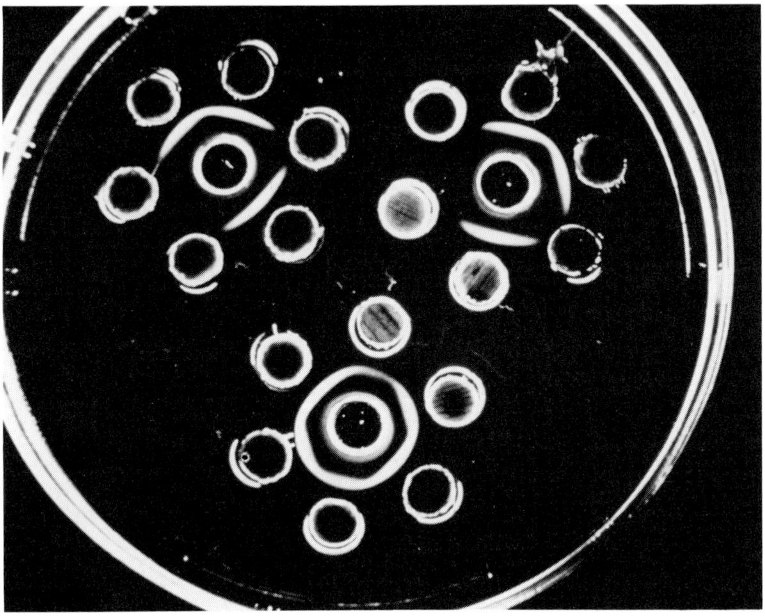

Fig. 23 Reactions in immunodiffusion seen 24-48 hours after preparation. The white lines are adjacent to an infected plant.

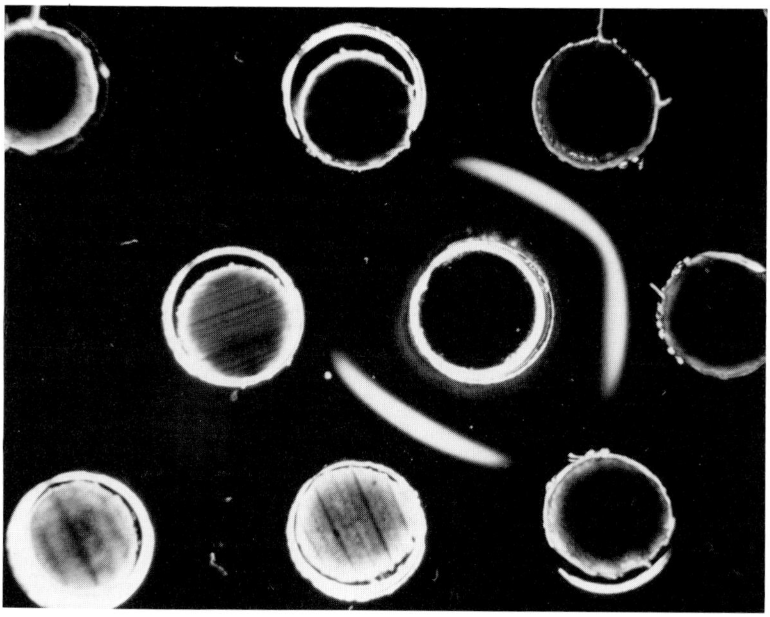

Fig. 24 Closeup of one six-well pattern. Reactions are (from top clockwise): Infected; healthy; infected; healthy; healthy; infected. Note that the healthy reactions show no line between the central well and the plant sample.

LOG SHEET

#	Identification of Plant Samples	Diagnosis
1		
2		
3		
4		
5		
6		
7		
8		
9		
10		
11		
12		
13		
14		
15		
16		
17		
18		

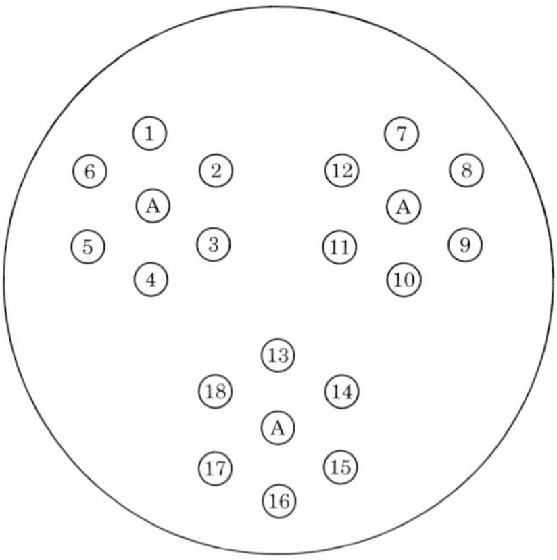

Test No. __________

Medium __________

Date __________

A = Antiserum

GLOSSARY

Antigen Usually a protein, in this case a virus which elicits an immune response in an immunized animal and reacts specifically with a corresponding antibody.

Antiserum Serum containing antibodies which specifically recognize a particular protein; in this case, a virus.

Chlorotic A yellowing of a normally green leaf.

Color break A disruption of the normal solid coloration of a pigmented flower.

Double antibody sandwich ELISA (DAS) A type of ELISA technique in which the virus is sandwiched between a specific antibody and the same antibody with an attached enzyme.

Dry pod propagation The mature seed pod is allowed to mature and burst naturally. The seeds are collected, sterilized, and placed on a tissue culture medium.

ELISA Enzyme-linked immunosorbent assay

Expressed symptoms Visible symptoms which the infected plant exhibits.

Green pod propagation The immature seed pod is harvested prematurely, surface sterilized, and the seeds are excised and placed on a sterile tissue culture medium.

Host range The range of plants which are susceptible to a particular virus — a diagnostic characteristic.

Immunodiffusion Also known as double radial diffusion, a serological technique in a petri plate with an agar medium. The virus and antibodies diffuse, meet and form a visible precipitin band.

Indicator plants Plants that are susceptible to a particular virus and when inoculated produce characteristic symptoms which indicate a viral infection.

Inoculate Deliberately infect a healthy plant with a virus. This can be done by rub-inoculating, or with aphids.

Leaf dip procedure An electron microscope technique in which the sap of a leaf is mixed in a buffer solution, placed in a tiny copper grid, then viewed under the EM.

Local infection Symptoms which are produced on an inoculated plant only on the inoculated leaves. The infection does not move beyond the inoculated leaf. Symptoms are usually discrete chlorotic spots.

Manual transmission Transmission of a virus by contaminated cutting tools, pots, or hands. No insect or other vector is involved.

Mosaic, Mottle Pattern of light and dark green.

Necrotic Dead-looking, also appears red or brown/black.

Negative controls Healthy plants which have not been inoculated.

Negative stain A stain in which virus particles seen in the electron microscope appear white against a dark background.

Positive controls Infected plants which give positive results in assays.

Serology The study of antigen and antibody interactions.

Substrate A compound which reacts specifically with an enzyme in ELISA, giving a color reaction.

Systemic infection A type of viral infection that continues to develop as the plant grows.

Titer Concentration of a virus

Vector An agent, usually an insect, which transmits a virus from one plant to another. This is a highly specific relationship between a particular vector and the virus it transmits.

Virus inclusion bodies Aggregates of millions of virus particles that form discrete, characteristic entities that can be seen under a light microscope.

SELECTED SUPPLIERS

There are many sources for seeds, chemicals, and diagnostic services besides those listed below. These represent a sampling of relevant, primary sources.

SEED SOURCES

W. Atlee Burpee Company
300 Park Avenue
Warminster, PA 18974
(215) 674-4900

J. L. Hudson, Seedsman
P. O. Box 1058
Redwood City, CA 94064

Includes weed sources for indicator plants

ANTISERA AND DIAGNOSTIC SERVICES

Ag-Dia
P. O. Box 1166
Mishawaka, Indiana 46544
Antisera — kits

ATCC (American Type Culture Collection)
12301 Parklawn Drive
Rockville, MD 20852
(800) 638-6597
Antisera

Critter Creek Laboratories
400 Critter Creek Road
Lincoln, CA 95648
ELISA test service

Orchid Virus Control Systems
G. C. Wisler
926 N. E. 7th Place
Gainesville, FL 32601
Consulting services

Orchis Laboratories
86 F Mason
Burdett, NY 14818
Immunodiffusion

Also consult:
State Departments of Agriculture

EQUIPMENT AND CHEMICALS

The Carborundum Company
Electro Minerals Division
Niagra Falls, NY
#500-RA

Carolina Biological Supply
Burlington, NC 27215
Stain kits for inclusion bodies

Eastman-Kodak Company
Rochester, NY 14650
8-quinolinol sulfate

Fisher Scientific
711 Forbes Avenue
Pittsburgh, PA 15219

Grafar Corporation
Detroit, MI
Gel punch

Safer Inc.
60 William Street
Suite G-20
Wellesley, MA 02181

Scientific Products:
Baxter Travenol Company
General Offices
1430 Waukegan Road
McGaw Park, Illinois 60085-6787

Sigma Chemical Company
P. O. Box 14508
St. Louis, MO 63178
(800) 325-3010, in U. S.
(314) 771-5750, outside U. S

Common U.S.A. Equivalents

Length:	1 inch . . . . =25.4001 millimeters
	1 millimeter . . . = 0.03937 inches
	1 foot . . . . = 0.304801 meters
	1 meter . . . . . . = 3.28083 feet
	1 yard . . . . = 0.914402 meters
	1 meter . . . . . . = 1.093611 yards
	1 mile . . . . = 1.609347 kilometers
Capacity: (Liquid)	1 quart . . . .= 0.94633 liters
	1 liter . . . .= 1.05671 quarts
	1 gallon . . . . . . = 3.78533 liters
	1 liter . . . . = 0.26418 gallons
Capacity: (Dry)	1 quart . . . . . . = 1.1012 liters
	1 liter . . . .= 0.9081 quarts
	1 peck . . . . = 8.810 liters
	1 liter . . . . = 0.11351 pecks
Mass: (Avoirdupois)	1 ounce . . . . . . = 28.350 grams
	1 gram . . . . = 0.035274 ounces
	1 pound . . . . . . = 0.45359 kilograms
	1 kilogram . . . . = 2.20462 pounds

COMMON MEASURES

3 teaspoons . . . . 1 tablespoon (1/2 ounce liquid)
2 tablespoons . . . . .1/8 cup (1 ounce liquid)
4 tablespoons . . . . .1/4 cup (2 ounces liquid)
5 tablespoons – 1 teaspoon . . . 1/3 cup
8 tablespoons . . . . .1/2 cup (4 ounces liquid)
10 tablespoons – 2 teaspoons . .2/3 cup
12 tablespoons . . . . .3/4 cup (6 ounces liquid)
16 tablespoons . . . . 1 cup (8 ounces liquid)
2 cups . . . . 1 pint (16 ounces liquid)
1 milliliter (ml) . . . . 1 cubic centimeter (cc)

APPENDIX

VIRUSES IN GESNERIADS

The 126 genera and more than 2,000 species in this family include African violets (*Saintpaulia* sp.), Gloxinia (*Sinningia* sp.), *Streptocarpus*, *Episcia*, and *Columnea* (44). The plants are valued for both their flowers and foliage. Although they are popular flowering and foliage ornamentals, viruses affecting this family have received little attention until recently.

Orchid growers often dabble in gesneriads — and vice versa. A familiarity with virus control in one group will help you with the other because the two families share many similarities. Both families are valued for the quality of their flowers, and with the gesneriads, their foliage as well. Both are plagued by some of the most stable viruses known, transmitted solely by the cultivation practices of hobbyists and commercial growers. In both families, extensive tests have shown that wild species are free of these viruses, which include TMV in gesneriads, and CyMV and ORSV in orchids.

Symptoms for both crops are often hard for the layman to detect, making it difficult to recognize virus infection and rogue infected plants. Viruses for both groups can be detected easily, however, through serological methods, particularly SDS-immunodiffusion.

Because of these similarities and the fact that most propagative techniques are vegetative, control measures are essentially the same. You should follow rules outlined for safe selection and sanitation practices of orchids with your gesneriads. This

TABLE 4. Incidence of tobacco mosaic virus (TMV) in cultivated and wild Gesneriaceae

	Grower[a]						
Genus	**A**	**B**	**C**	**D**	**E**	**F**	**Wild[b]**
Achimenes	2/2	19/26	—	—	—	—	0/23
*Codonanthe**[c]	0/1	6/11	3/16	—	—	—	0/5
*Columnia**	16/42	23/97	—	10/17	—	1/5	0/16
*Episcia**	0/38	0/43	—	2/2	—	—	0/5
Gloxinia	0/1	—	—	1/1	1/1	—	0/5
Kohleria	—	7/13	—	—	1/1	3/4	0/6
*Nematanthus**	2/19	1/31	1/19	0/1	—	—	0/5
Sinningia	1/3	—	—	—	2/4	—	0/5
*Smithiantha**	—	—	—	—	8/9	—	0/3
Miscellaneous[d]	4/40	6/83	—	0/1	1/1	—	—

[a] Ratios are number of infected plants per total number tested; — denotes no samples tested. Results are based on serological and/or electron microscopy assays.

[b] All samples represent herbarium specimens.

[c] * denotes not previously reported as susceptible to TMV.

[d] This category includes the following: *Aeschynanthus** 5/64, *Streptocarpus** 3/22, *Eucodonia* 0/3, *Niphaea* 0/1, *Chirita** 1/2, *Chrysothemis* 0/1, *Saintpaulia* 0/16, *Drymonia* 0/6, ×*Achimenantha* 1/3, ×*Coltrichantha** 1/2, and ×*Codonatanthus* 0/5.

includes the use of sterile tools, pots and media.

Scientists at the University of Florida (44, 45) have attempted to determine the occurrence and importance of tobacco mosaic virus (TMV) in gesneriads. TMV is the "type-member" of the tobamovirus group and a close relative of ORSV that affects orchids. Results show that the TMV is highly prevalent within cultivated gesneriads (Table 4), found in 14 of 20 genera surveyed, including ten genera not previously reported.

The highest incidence of virus — from 31 to 89 percent of samples — were found in the genera *Achimenes*, *Codonanthe*, *Columnea*, *Kohleria* and *Smithiantha*. Lower incidences, from two to eight percent, were found in the genera *Aeschynanthus*, *Episcia*, and *Nematanthus*. TMV was found in all six sampled collections of cultivated Gesneriads, ranging from 11-81 percent. No infection of TMV was found in any sample of *Saintpaulia*.

As with wild orchids, TMV was not found to infect wild gesneriads (Table 4). The two strains of TMV detected in cultivated gesneriads (U_1 and U_2) also exist in smoking tobacco, unlike ORSV in orchids. Since they are not found in wild specimens, it is likely the viruses were introduced into the family from tobacco and carried along by vegetative propagation.

Since smoking tobacco is a source of infection to the gesneriads, it is wise not to smoke while working on these plants. Also take care to thoroughly scrub your hands and fingernails after you smoke before entering your greenhouse.

Other viruses besides those already mentioned have been reported in gesneriads. These include a bacilliform-shaped virus in African violet (11); tomato spotted wilt; tomato ringspot, and dahlia ringspot (45). Little is known about their incidence in the industry, however.

One unidentified virus is worth mentioning, though, in light of the orchid virus situation. A study revealed a virus belonging to the potexvirus group, which is the same as cymbidium mosaic, in four of nine *Smithiantha* samples. The particle measurements fit the size for protexviruses but serological studies comparing this virus to other members of that group, including CyMV, were unsuccessful. Although the exact identity of this virus is unknown, it is known to be manually transmitted by vegetative propagation.

A word of cautious reassurance to hobbyists who grow both gesneriads and orchids: it is unlikely that a virus infecting orchids would infect gesneriads, and vice versa. No definitive tests have been conducted to prove this, however.

VIRUSES IN GESNERIADS AND ORCHIDACEAE:
Similarities and Differences

Similarities

1. Both plant families are plagued by some very stable, manually transmitted viruses that can be controlled simply by sanitation procedures and roguing of infected plants.
2. These viruses are not found in wild plants and appear to be a product of cultivation practices.
3. Serological techniques are extremely useful in identifying infected plants.
4. Both plant groups have viruses which are closely related.

Differences

1. ORSV that infects orchids, and TMV which infects gesneriads, are closely related, yet they are distinct viruses.
2. TMV, found in smoking tobacco, is the same virus as that found in gesneriads. TMV does not infect orchids systemically, but only produces local lesions on inoculated orchid leaves.
3. CyMV is the most important virus of orchids. CyMV belongs to the potexvirus group. Another member of that group also infects gesneriads, but little is known about it.

REFERENCES

1. Allison, A. V. 1981. Enyzme-linked immunosorbent assay (ELISA) as a rapid method for screening for the presence of cymbidium mosaic and odontoglossum ringspot virus. *Phytopathology* 71, No.1.

2. Bodnaruk, W. H., G. R. Hennen, F. W. Zettler, and T. J. Sheehan. 1979. Incidence of cymbidium mosaic and odontoglossum ringspot viruses in wild and cultivated orchids of the Cattleya alliance surveyed in Florida. *American Orchid Society* 48: 26-27.

3. Bos, L. 1970. Bean yellow mosaic virus. *CMI/AAB.* Description of plant viruses, No. 40.

4. Burnett, H. C. 1965. Orchid Diseases. Florida Dept. of Agric. and Consumer Serv., Div. of Plant Ind. Bulletin 10:50-60.

5. Burnett, H. C. 1967. Orchid mericlones found infected with cymbidium mosaic virus. Fla. Dept. Agric. and Consumer Serv., Div. Plant Ind. Pathology Circular No. 56.

6. Burnett, H. C. 1984. Guignardia leafspot of *Ascocentrum*,. *Ascocenda*, and species and hybrids of vanda orchids. Fla. Dept. Agric. and Consumer Serv., Div. Plant Ind. Circular No. 258.

7. Carter, H. July 1980. Florida's wildlife detectives. *Florida Sportsman.* pp. 75-77.

8. Chang, M. U., K. Arai, Y. Doi, and K. Yora. 1976. Morphology and intracellular appearance of orchid fleck virus. *Annual Phytopathology Society of Japan.* 42:156-167.

9. Chang, M. U., Y. Doi, and K. Yora. 1978. Cymbidium

mild mosaic virus, isolation of cymbidium mild mosaic virus. *Korean Journal Plant Protection.* 17:131-138.

10. Christie R. G., N.-J. Ko, and F. W. Zettler. 1986. Light microscopic techniques for diagnosis of orchid virus diseases. *American Orchid Society Bulletin.* 55:996-1007.

11. Ciampor, F. and DoKoupil, N. 1974. Electron microscopic demonstration of a bacilliform virus in *Saintpaulia ionantha. Acta Virol.* 18: 355-357.

12. Clark, M. F. 1981. Immunosorbent assays in plant pathology. *Ann. Rev. Phytopathol.* 19:83-106.

13. Clark M. F. and A. N. Adams. 1977. Characteristics of the microplate method of the enzyme-linked immunosorbent assay (ELISA) for the detection of plant viruses. *J. Gen. Virol.* 34: 475-485.

14. Doi, Y., M. U. Chang, and K. Yora. 1977. Orchid fleck virus. CMI/AAB. Description of plant viruses. No. 183.

15. Elliott, M. S., F. W. Zettler, T. J. Sheehan, and G. C. Wisler. 1987. A virus disease of *Dendrobium* in Florida. *American Orchid Society.* 55:1052-3.

16. Francki. R. I. B. 1966. Isolation, purification and some properties of two viruses from cultivated cymbidium orchids. *Aust. J. Biol. Sci.* 19:555-564.

17. Goff, L. M., and M. K. Corbett. 1977. Association of tomato ringspot virus with a chlorotic leaf streak of cymbidium orchids. *Phytopathology.* 67:1096-1100.

18. Hakkaart, IR. F. A. 1980. Virusziekten in *Cymbidium. Vakbladvoor de bloemisterif.* 21:52-55.

19. Hamon, A. B. 1986. Orchid Pests. In *Handbook of Orchid Pests and Diseases.* American Orchid Society. pp. 50-63.

20. Harris, K. F. 1981. Arthropod and nematode vectors of plant viruses. *Annual Rev. Phytopathol.* 19:391-426.

21. Hollings, M., and O. M. Stone. 1977. Cymbidium ringspot virus. CMI/AAB. Description of plant viruses, No. 178.

22. Inouye, N. 1969. Cucumber mosaic virus isolated from *Dendrobium. Nogaku Kenkyu.* 53:49-60.

23. Ishii, M. 1974. Partial elimination of virus from doubly infected orchids by meristem explant culture. Third International Symposium on Virus Diseases in Ornamental Plants. *International Society for Hort. Sci. Acta Horticulturae.* Number 36, pp. 229-233.

24. Ko, N.-J., F. W. Zettler, and G. C. Wisler. 1985. A simplified bioassay technique for cymbidium mosaic and odontoglossum ringspot viruses. *American Orchid Society Bulletin.* 54:1080.

25. Lawson, R. H. 1967. Chemical inactivation of cymbidium mosaic and odontoglossum ringspot viruses. *American Orchid Society Bulletin.* 36:998-1001.

26. Lawson, R. H. and M. Brannigan. 1986. Virus diseases of orchids. In *Handbook of orchid pests and diseases.* American Orchid Society. pp. 2-49.

27. Lawson, R. H. and R. K. Horst. July 1984. Tests help prevent viruses in orchids. *Greenhouse Manager.* pp. 102-108.

28. Lesemann, D. E. 1977. Long filamentous virus-like particles associated with vein necrosis of *Dendrobium phalaenopsis. Phytopath. Z.* 89: 330-339.

29. Morel, G. M. 1960. Producing virus-free cymbidiums. *American Orchid Society Bulletin.* 29:495-497.

30. Northen, R. T. 1970. *Home orchid growing.* Van Nostrand Reinhold Company, New York. 374 pp.

31. Schuster, G. 1979. On some interactions of 1-B-D-ribofuranosyl-1,2,4-triazole-3-carboxamide (Virazole) and plant hormones in virus infected plants. *Phytopath Z.* 94: 72-79.

32. Sheehan, T. J. 1980. Orchids. In *Introduction to Floriculture,* Roy A. Larson (ed.), Chapter 5. Academic Press, Raleigh, N. C.

33. Short, D. E., and G. W. Simone. 1986. Pesticides. In *Handbook of orchid pests and diseases.* American Orchid

Society. pp. 64-70.

34. Sproles, E. T. III. 1983. An effort to eradicate cymbidium mosaic and odontoglossum ringspot viruses from a private collection — results of a four-year study. *American Orchid Society Bulletin*. 52:261-265.

35. Sproles, E. T. III. 1987. Maintaining a virus-free collection: a four-year follow-up. *American Orchid Society Bulletin*. 56:1054-6.

36. Wisler, G. C. 1981. Development of serological techniques designed to index cymbidium mosaic and odontoglossum ringspot viruses in commercial orchids. (M. S. Thesis), University of Florida.

37. Wisler, G. C., F. W. Zettler, and L. Mu. 1987. Viruses infecting *Vanilla* in French Polynesia. *American Orchid Society Bulletin*. 56:361-387.

38. Wisler, G. C., F. W. Zettler, and D. E. Purcifull. 1982. A serodiagnostic technique for detecting cymbidium mosaic and odontoglossum ringspot viruses. *Phytopathology*. 72:835-837.

39. Wisler, G. C., F. W. Zettler, and T. J. Sheehan. 1979. Relative incidence of cymbidium mosaic virus and odontoglossum ringspot viruses in several genera of wild and cultivated orchids. *Proc. Fla. State. Hort. Soc.* 92:339-340.

40. Wisler, G. C., F. W. Zettler, and T. J. Sheehan. 1983. A diagnostic technique developed for orchid growers to detect cymbidium mosaic and odontoglossum ringspot viruses. *American Orchid Society Bulletin*. 52:255-61.

41. Wisler, G. C., F. W. Zettler, and T. J. Sheehan. 1986. Common questions and misconceptions concerning orchid viruses. *American Orchid Society Bulletin*. 55:472-479.

42. Yuen, C. K. K. H., H. Kanemoto, and M. Ishii. 1979. Transmission of cymbidium mosaic virus through seed propagation in *Dendrobium*. *American Orchid Society Bulletin*. 48:1245-1247.

43. Zettler, F. W., G. R. Hennen, W. H. Bodnaruk, Jr., H. T. Clifford and T. J. Sheehan. 1978. Wild and cultivated

orchids surveyed in Florida for the cymbidium and odontoglossum ringspot viruses. *Plant Dis. Reptr.* 62:949-952.

44. Zettler, F. W., and J. Nagel. 1983. Tobacco mosaic virus infections of gesneriads. *The Gloxinian.* 33: 22-25.

45. Zettler, F. W. and J. Nagel. 1983. Infection of cultivated gesneriads by two strains of tobacco mosaic virus. *Plant Disease.* 67: 1123-1125.

46. Zettler, F. W., G. C. Wisler, M. S. Elliott, and N.-J. Ko. 1987. Some new, potentially significant viruses of orchids and their probable means of transmission. *American Orchid Society Bulletin.* 55:1045-1051.

47. Zettler, F. W., G. C. Wisler, T. J. Sheehan, N.-J. Ko, and A. E. Logan. 1984. Some basics and a perspective for orchid growers. *American Orchid Society Bulletin.* 53:505-511.

INDEX

ABOUT THE AUTHOR

Gail C. Wisler, who is a plant pathologist, has extensive experience with plant virus identification. She was responsible for virus diagnosis of citrus, orchids, lettuce, and a variety of foliage, ornamental, vegetable, and fruit crops for 10 years at the Florida Department of Agriculture's Division of Plant Industry. She then headed a monoclonal antibody laboratory for plant viruses at the University of Florida.

The author of numerous published articles about orchid viruses, Ms. Wisler is frequently called on to present talks and workshops on the subject. She has a B. S. degree in biology from the College of William and Mary, Williamsburg, Virginia. She holds an M. S. degree in plant pathology from the University of Florida, where she concentrated her studies on a serological technique designed to index viruses and aid in the control of cymbidium mosaic and odontoglossum ringspot viruses of orchids. She is pursuing a Ph.D in plant pathology at UF.

Ms. Wisler operates Orchid Virus Control Systems, an orchid virus testing and consulting business in Gainesville, Florida.

Ordering information

How to Control Orchid Viruses: The Complete Guidebook,

by Gail Wisler.

Single copy price -- $22.45 (includes 6 percent sales tax and $1.50 handling)

Check, money order, or Master Card/VISA payment accepted. Please enclose credit card number and expiration date.

Book Order Department
Maupin House Publishing
P.O. Box 90148
Gainesville, FL 32607
(904) 336-9290

Bulk discount rates available on request.